Vibrations and Waves and Electrical Calculations:

A Physics Book for High Schools and Colleges

By

Kingsley Augustine

TABLE OF CONTENT

CHAPTER 1 SIMPLE HARMONIC MOTION 3

CHAPTER 2 ENERGY IN SIMPLE HARMONIC MOTION 18

CHAPTER 3 CIRCULAR MOTION 25

CHAPTER 4 WAVE MOTION 32

CHAPTER 5 ECHOES 42

CHAPTER 6 BEAT 47

CHAPTER 7 VIBRATION OF AIR COLUMN IN PIPES 49

CHAPTER 8 MODES OF VIBRATION OF A STRETCHED STRING 55

CHAPTER 9 CHARACTERISTICS OF SOUND – THE PITCH 64

CHAPTER 10 DOPPLER EFFECTS IN SOUND 67

CHAPTER 11 ELECTRIC CURRENT 74

CHAPTER 12 RESISTORS IN CIRCUITS 79

CHAPTER 13 DIVISION OF CURRENT AND VOLTAGES BETWEEN RESISTORS IN CIRCUITS 84

CHAPTER 14 GENERAL CALCULATIONS IN ELECTRIC CIRCUITS 93

CHAPTER 15 ELECTRICAL ENERGY 98

CHAPTER 16 BUYING OF ELECTRICAL ENERGY 103

CHAPTER 17 MEASUREMENT OF RESISTANCE 107

CHAPTER 18 LAWS OF ELECTROLYSIS 112

CHAPTER 19 CONVERSION OF GALVANOMETER TO AMMETER AND VOLTMETER 119

CHAPTER 20 ALTERNATING CURRENT (A.C) CIRCUIT 125

CHAPTER 21 RESISTOR, INDUCTOR AND CAPACITOR (R-L-C) CIRCUIT IN SERIES 129

ANSWERS TO EXERCISES 146

CHAPTER 1
SIMPLE HARMONIC MOTION

Simple harmonic motion (SHM) is defined as the motion of a body whose acceleration is directly proportional to the displacement from a fixed point and is always directed towards that fixed point.

Three common examples of bodies in simple harmonic motion are:
1. A swinging pendulum bob
2. A displaced mass hanging from a spiral spring
3. A loaded test tube depressed and released in a liquid

Velocity and Acceleration of Simple Harmonic Motion (SHM)

The maximum velocity obtained by a body in SHM is given by:

$$v = \omega A$$

where ω = angular velocity in radians/second (rad/sec), and A = maximum displacement of the body (i.e. the amplitude).

However, at a point which is at a distance of x, from the center of motion (equilibrium or fixed point), the velocity is given by:

$$v = \omega \sqrt{A^2 - x^2}$$

The velocity is zero at the ends of motion, and maximum at the center of motion.
The acceleration, a, for a body performing SHM is given by:

$$a = \omega^2 x$$

where x is the distance of the body from the center of the motion.
The acceleration is zero at the center of motion and maximum at the ends of motion.
It should be noted that this acceleration is negative.

Period and Frequency of SHM

1. **Period**: This is the time taken to complete one cycle or oscillation or vibration.
In terms of number of oscillation/cycle, and time taken to complete the oscillation, period, T, is given by:

$$T = \frac{\text{Time taken}}{\text{Number of oscillations /cycles}}$$

2. **Frequency**: This is the number of cycles or oscillations completed in one second.
In terms of number of oscillation/cycle, and time taken to complete the oscillation, frequency, f, is given by:

$$f = \frac{\text{Number of oscillations /cycles}}{\text{time taken}}$$

The unit of period is second while the unit of frequency is Hertz (Hz).

Generally, for any object performing SHM, the following formulas apply:

- Angular velocity, $\omega = 2\pi f$ or $\omega = \dfrac{2\pi}{T}$
- Period, $T = \dfrac{2\pi}{\omega}$ or $T = \dfrac{1}{f}$
- Frequency, $f = \dfrac{\omega}{2\pi}$ or $f = \dfrac{1}{T}$

Period of Simple Pendulum

Another formula that can be used to calculate the period of a simple pendulum is given by:

$$T = 2\pi \sqrt{\dfrac{l}{g}}$$

where l is the length of the pendulum.

The equation above shows that the angular velocity for a simple pendulum is given by:

$$\omega = \sqrt{\dfrac{g}{l}} \quad \text{(when } T = \dfrac{2\pi}{\omega} \text{ is compared with } T = 2\pi \sqrt{\dfrac{l}{g}}\text{)}$$

When a simple pendulum of length l_1 and period T_1 is compared to another pendulum of length l_2 and period T_2, then the relation between them is given by:

$$\dfrac{T_1}{T_2} = \sqrt{\dfrac{l_1}{l_2}}$$

Or, $\left(\dfrac{T_1}{T_2}\right)^2 = \dfrac{l_1}{l_2}$

Period of a Spiral Spring

A formula that can be used to calculate the period of a loaded spiral spring in SHM is:

$$T = 2\pi \sqrt{\dfrac{m}{k}} \quad \text{or} \quad T = 2\pi \sqrt{\dfrac{e}{g}}$$

where m is the mass attached to the spiral spring, k is the elastic constant of the spring, and e is the extension produced by the mass attached to the spring.

The equation above shows that the angular velocity for a spiral spring is given by:

$$\omega = \sqrt{\dfrac{k}{m}} \quad \text{or} \quad \omega = \sqrt{\dfrac{g}{e}} \quad \text{(when } T = \dfrac{2\pi}{\omega} \text{ is compared with } T = 2\pi \sqrt{\dfrac{m}{k}} \text{ and } T = 2\pi \sqrt{\dfrac{e}{g}}\text{)}$$

When a spiral spring has mass, m_1, attached to it and it has period T_1, and it is compared to a similar spiral spring with mass, m_2, attached to it and has period T_2, then the relation between them is given by:

$$\dfrac{T_1}{T_2} = \sqrt{\dfrac{m_1}{m_2}}$$

Or, $\left(\dfrac{T_1}{T_2}\right)^2 = \dfrac{m_1}{m_2}$

Similarly, when the extension, e_1, on a spiral spring of period T_1. is compared to a similar spiral spring with extension, e_2, and period, T_2, then the relation between them is given by:

$\dfrac{T_1}{T_2} = \sqrt{\dfrac{e_1}{e_2}}$

Or, $\left(\dfrac{T_1}{T_2}\right)^2 = \dfrac{e_1}{e_2}$

Period of a Loaded Test Tube in a Liquid

For a loaded test tube depressed in a liquid and allowed to perform SHM, if its mass is m, cross section area is A, and density of the liquid is ρ, then the period is given by:

$T = 2\pi \sqrt{\dfrac{m}{A\rho g}}$

The equation above shows that the angular velocity for a loaded test tube is given by:

$\omega = \sqrt{\dfrac{A\rho g}{m}}$ (when $T = \dfrac{2\pi}{\omega}$ is compared with $T = 2\pi \sqrt{\dfrac{m}{A\rho g}}$)

When we compare two similar loaded test tubes in the same liquid, then the relationship connecting their periods and masses is given by:

$\dfrac{T_1}{T_2} = \sqrt{\dfrac{m_1}{m_2}}$

Or, $\left(\dfrac{T_1}{T_2}\right)^2 = \dfrac{m_1}{m_2}$

Equations of Simple Harmonic Motion

The generalized equation of the position of a body in simple harmonic motion as a function of time is given by:

$x = A \cos(\omega t + \phi)$

where t is the time in seconds, ω is the angular velocity/frequency, A is the amplitude, and ϕ is the phase difference/shift in radians. When the equation above is differentiated, it gives the equation of the velocity of a body in SHM as follows:

$v = -A\omega \sin(\omega t + \phi)$

When the equation for velocity is differentiated, it gives the equation for acceletation of a body in SHM as follows:

$a = -A\omega^2 \cos(\omega t + \phi)$

Examples

1. A body performing simple harmonic motion has a maximum displacement from the center of motion to be 0.2m. If its angular velocity is 6rad/sec, calculate:
(a) the period
(b) the frequency
(c) the maximum velocity
(d) the acceleration at the center and at the end of motion
(e) the velocity of the body at a point 0.12m from the center of motion
(Take π = 3.142)

Solution

(a) $T = \dfrac{2\pi}{\omega}$

$= \dfrac{2 \times 3.142}{6}$

$= 1.05$ sec

(b) $f = \dfrac{\omega}{2\pi}$

$= \dfrac{6}{2 \times 3.142}$

$= 0.95$ Hz

(c) $v = \omega A$

$= 6 \times 0.2$

$v = 1.2$ m/s

(d) The acceleration at the center is zero.
The acceleration at the end of motion is maximum, and is given by:

$a = \omega^2 x$

$= 6^2 \times 0.2$

$= 36 \times 0.2$

$= 7.2$ ms^{-2}

(e) This velocity at a certain point is given by:

$v = \omega\sqrt{A^2 - x^2}$

$= 6 \times \sqrt{0.2^2 - 0.12^2}$

$= 6 \times \sqrt{0.04 - 0.0144}$

$= 6 \times \sqrt{0.0256}$

$= 6 \times 0.16$

$= 0.96$ m/s

2. A simple pendulum of length 60cm oscillates with amplitude of 0.05m. Calculate:
(a) the period of oscillation
(b) the maximum velocity of the motion   (Take g = 10m/s^2)
Solution

(a) $T = 2\pi\sqrt{\dfrac{l}{g}}$

$= 2 \times 3.142 \times \sqrt{\dfrac{0.6}{10}}$ (Note that 60cm = $(\dfrac{60}{100})$m = 0.6m)

$= 6.284 \times \sqrt{0.06}$

$= 1.54$ sec

(b) v = ωA

But, $\omega = \dfrac{2\pi}{T}$

$= \dfrac{2 \times 3.142}{1.54}$

ω = 4.0805 radsec^{-1}

Hence, v = ωA

$= 4.0805 \times 0.05$

$= 0.204$ ms^{-1}

3. An object performing simple harmonic motion has an angular velocity of 5rad/sec. If the amplitude of the motion is 30cm, calculate the velocity of the object at a point:
(a) 20cm from the equilibrium position
(b) 12cm from the end of motion
Solution

(a) The amplitude, A = 30cm = $\dfrac{30}{100}$ = 0.3m

$x = \dfrac{20}{100} = 0.2$m

The velocity at a certain point is given by:

$v = \omega\sqrt{A^2 - x^2}$

$= 5 \times \sqrt{0.3^2 - 0.2^2}$

$= 5 \times \sqrt{0.09 - 0.04}$

$= 5 \times \sqrt{0.05}$

$= 5 \times 0.2236$

$= 1.118$ m/s

(b) In this case, the distance of the object from the center of motion is given by:

$x = 0.3 - 0.12$ (Note that 12cm = 0.12m)

$x = 0.18$

Hence the velocity at 0.12m from end of motion (i.e. 0.18m from center) is given by:

$v = \omega\sqrt{A^2 - x^2}$

$= 5 \times \sqrt{0.3^2 - 0.18^2}$

$= 5 \times \sqrt{0.09 - 0.0324}$

$= 5 \times \sqrt{0.0576}$

$= 5 \times 0.24$

$= 1.2 \text{m/s}$

4. The acceleration of body in simple harmonic motion is 64 times its displacement in meters. Determine the frequency of the motion.

Solution

Displacement is x. Hence, acceleration is $64x$. Let us now substitute each value into the formula for acceleration as follows:

$a = \omega^2 A$

$64x = \omega^2 x$

$64 = \omega^2$ (x has cancelled out)

$\omega = \sqrt{64}$

$\omega = 8$

But, $f = \dfrac{\omega}{2\pi}$

$= \dfrac{8}{2 \times 3.142}$

$f = 1.27 \text{Hz}$

5. The period of a simple pendulum P is 3sec. What is the period of a simple pendulum Q which makes 250 vibrations in the time it takes P to make 200 vibrations.

Solution

Recall that: $T = \dfrac{\text{Time taken}}{\text{Number of oscillations /cycles}}$

Substituting values for pendulum P gives:

$3 = \dfrac{t}{200}$ (t is the time taken by pendulum P)

$t = 3 \times 200$

$t = 600 \text{sec}$

This duration of 600 seconds taken by P is also the same time taken by Q. Hence, we now substitute appropriate values for Q in order to obtain its period.

$$T = \frac{\text{Time taken}}{\text{Number of oscillations /cycles}}$$

$$= \frac{600}{250}$$

T = 2.4

Therefore, the period of pendulum Q is 2.4 seconds.

6. An object moving with simple harmonic motion has an amplitude of 8cm and a frequency of 60Hz. Calculate:

(a) the period of oscillation

(b) the velocity at the middle and end of oscillation

Solution

(a) The angular velocity is given by:

$\omega = 2\pi f$

$= 2 \times 3.142 \times 60$

$\omega = 377.04$ rad/sec

The period is given by:

$$T = \frac{2\pi}{\omega}$$

$$= \frac{2 \times 3.142}{377.04}$$

$$= 0.0167 \text{sec}$$

(b) The velocity at the middle of oscillation is the maximum velocity given by:

$v = \omega A$

$= 377.04 \times 0.08$ (Note that 8cm = 0.08m)

$= 30.16 \text{ms}^{-1}$

7. The period of a simple pendulum is 4.4sec. When the length of the pendulum is reduced by 1m, the period is 3.8sec. Determine:

(a) the original length of the pendulum

(b) the value of the acceleration due to gravity of the place.

Solution

(a) The formula for comparing the periods of two simple pendulums is given by:

$$\left(\frac{T_1}{T_2}\right)^2 = \frac{l_1}{l_2}$$

The original length is l_1. Hence, the new length is $l_2 = l_1 - 1$ (since original length was reduced by 1m to obtain the new final length). Substituting into the formula above gives:

$$\left(\frac{T_1}{T_2}\right)^2 = \frac{l_1}{l_2}$$

$$\left(\frac{4.4}{3.8}\right)^2 = \frac{l_1}{l_1 - 1}$$

$$1.3407 = \frac{l_1}{l_1 - 1}$$

$$1.3407(l_1 - 1) = l_1$$

$$1.3407 l_1 - 1.3407 = l_1$$

$$1.3407 l_1 - l_1 = 1.3407$$

$$0.3407 l_1 = 1.3407$$

$$l_1 = \frac{1.3407}{0.3407}$$

$$l_1 = 3.94$$

Therefore, the original length of the pendulum is 3.94m.

(b) Let us use the formula for the period to obtain g as follows.

$$T = 2\pi \sqrt{\frac{l}{g}}$$

$$4.4 = 2 \times 3.142 \times \sqrt{\frac{3.94}{g}} \quad \text{(use the period of the original pendulum)}$$

$$4.4 = 6.284 \sqrt{\frac{3.94}{g}}$$

Square both sides of the equation to obtain:

$$4.4^2 = 6.284^2 \left(\frac{3.94}{g}\right) \quad \text{(Note that the root sign has gone)}$$

$$19.36 = \frac{155.585}{g}$$

$$g = \frac{155.585}{19.36}$$

$$g = 8.04$$

Therefore the value of the acceleration due to gravity is 8.04ms^{-2}.

8. The period of a simple pendulum is 10sec. Calculate its period when its length is tripled.
Solution
The formula for comparing the periods of two simple pendulums is given by:

$$\left(\frac{T_1}{T_2}\right)^2 = \frac{l_1}{l_2}$$

The original length is l_1. Hence, the new length is $l_2 = 3 l_1$ (since original length was tripled).
Substituting into the formula above gives:

$$\left(\frac{T_1}{T_2}\right)^2 = \frac{l_1}{l_2}$$

$$\left(\frac{10}{T_2}\right)^2 = \frac{l_1}{3l_1}$$

$$\frac{100}{T_2^2} = \frac{1}{3} \quad (l_1 \text{ has cancelled out})$$

$$= 3 \times 100$$

$$T_2 = \sqrt{300}$$

$$T_2 = 17.32$$

Therefore, the period is 17.32 seconds.

This shows that the new period can be obtained by applying the formula:

$$T_2 = \sqrt{factor\ of\ multiplication} \times T_1$$

In our example above: $T_2 = \sqrt{3} \times 10$

$$= 1.732 \times 10$$

$$= 17.32 \text{ (as obtained above)}$$

9. The period of a spiral spring is 5sec when the mass attached to it is 20g. What is its period when this mass is replaced with an 80g mass?

Solution

The formula for comparing the periods of a spiral spring is given by:

$$\left(\frac{T_1}{T_2}\right)^2 = \frac{m_1}{m_2}$$

Substituting values into the formula above gives:

$$\left(\frac{5}{T_2}\right)^2 = \frac{20}{80}$$

$$\frac{25}{T_2^2} = \frac{1}{4}$$

$$T_2 = 4 \times 25$$

$$T_2 = \sqrt{100}$$

$$T_2 = 10$$

Therefore, the period is 10 seconds.

10. A simple pendulum of length 40cm performs simple harmonic motion. Determine the angular velocity of the pendulum. (g = 10m/s²)

Solution

For a simple pendulum, the angular velocity can be obtained as follows:

$$\omega = \sqrt{\frac{g}{l}}$$

$$= \sqrt{\frac{10}{0.4}} \quad \text{(Note that 40cm = 0.4m)}$$

$= \sqrt{25}$

$\omega = 5 \text{rad/sec}$

11. A spiral spring has a mass of 20g attached to it, and this extends the spring by 5cm. The spring is pulled down a distance of 10cm and allowed to perform simple harmonic motion. Calculate:

(a) the angular velocity of the motion

(b) the velocity of the spring at the point 8cm from the mean position.

$(g = 10 \text{m/s}^2)$

Solution

(a) For a spiral spring, the angular velocity can be obtained as follows:

$\omega = \sqrt{\dfrac{g}{e}}$

$= \sqrt{\dfrac{10}{0.05}}$ (Note that 5cm = 0.05m)

$= \sqrt{200}$

$\omega = 14.14 \text{rad/sec}$

(b) $v = \omega\sqrt{A^2 - x^2}$

where A = 10cm = 0.1m and x = 8cm = 0.08m. Substituting these values into the equation above gives:

$v = 14.14\sqrt{0.1^2 - 0.08^2}$

$= 14.14\sqrt{0.0036}$

$= 14.14 \times 0.06$

$v = 0.848 \text{m/s}$

12. The equation of the displacement in meters, of a body in simple harmonic motion is given by:

$x = 6 \cos(\pi t + \dfrac{\pi}{3})$

where t is in seconds. Determine:

(a) the amplitude, frequency and period of motion

(b) the equations of the velocity and acceleration of the body

(c) position, velocity and acceleration of the body at time t = 1 second

Solution

(a) The given equation: $x = 6 \cos(\pi t + \dfrac{\pi}{3})$

The general equation: $x = A \cos(\omega t + \phi)$

Comparing the two equations above shows that the amplitude, A = 6m.

It also shows that the angular velocity, $\omega = \pi$

Hence, $2\pi f = \pi$ (Note that $\omega = 2\pi f$)

Therefore, $f = \dfrac{\pi}{2\pi}$ (When both sides are divided by 2π)

$f = \dfrac{1}{2}$ (π cancels out)

f = 0.5Hz

The period is given by:

$T = \dfrac{1}{f}$

$= \dfrac{1}{0.5}$

T = 2 seconds

(b) $x = 6 \cos(\pi t + \dfrac{\pi}{3})$

$x = A \cos(\omega t + \phi)$

Comparing the equations above shows that: A = 6, $\omega = \pi$, and $\phi = \dfrac{\pi}{3}$

The general equation for the velocity of a body in SHM is given by:

$v = -A\omega \sin(\omega t + \phi)$

Substituting the appropriate values into the equation above gives the equation for the velocity of the body as follows:

$v = -6\pi \sin(\pi t + \dfrac{\pi}{3})$

The general equation for the acceleration of a body in SHM is given by:

$a = -A\omega^2 \cos(\omega t + \phi)$

Substituting the appropriate values into the equation above gives the equation for the acceleration of the body as follows:

$a = -6\pi^2 \cos(\pi t + \dfrac{\pi}{3})$

(c) The equation for the position of the body is given by:

$x = 6 \cos(\pi t + \dfrac{\pi}{3})$

At time t = 1 second, the position is obtained by substituting 1 for t in the equation above. This gives:

$x = 6 \cos(\pi(1) + \dfrac{\pi}{3})$

$= 6 \cos(\pi + \dfrac{\pi}{3})$

$= 6 \cos \frac{4\pi}{3}$

$= 6 \times (-0.5)$ (Note that $\cos \frac{4\pi}{3} = -0.5$, where $\frac{4\pi}{3}$ is in radians not degree)

$x = -3m$

The equation for the velocity of the body is given by:

$v = -6\pi \sin (\pi t + \frac{\pi}{3})$

At time t = 1 second, the velocity is obtained by substituting 1 for t in the equation above. This gives:

$v = -6\pi \sin (\pi(1) + \frac{\pi}{3})$

$= -6\pi \sin (\pi + \frac{\pi}{3})$

$= -6\pi \sin \frac{4\pi}{3}$

$= -6 \times \pi \times (-0.866)$ (Note that $\sin \frac{4\pi}{3} = -0.866$, where $\frac{4\pi}{3}$ is in radians)

$v = 5.196 \times 3.142$ (Note that $\pi = 3.142$)

$v = 16.33 m/s$

The equation for the acceleration of the body is given by:

$a = -6\pi^2 \cos (\pi t + \frac{\pi}{3})$

At time t = 1 second, the acceleration is obtained by substituting 1 for t in the equation above. This gives:

$a = -6\pi^2 \cos (\pi(1) + \frac{\pi}{3})$

$= -6\pi^2 \cos (\pi + \frac{\pi}{3})$

$= -6\pi^2 \cos \frac{4\pi}{3}$

$= -6 \times \pi \times \pi \times (-0.5)$

$a = 3 \times 3.142 \times 3.142$

$a = 29.62 m/s^2$

Note that in working with angles that are in radians, your calculator has to be set to radians not degrees. However, the angles in radians in the problem above can be converted to angles in degrees by simply substituting 180 for π, because 3.142 radians = 180 degrees (i.e. π radians = 180°). For example:

$\cos \frac{4\pi}{3} = \cos(\frac{4 \times 180}{3}) = \cos 240 = -0.5$

Hence, $\cos \frac{4\pi}{3} = -0.5$ (where $\frac{4\pi}{3}$ is in radians)

Or, $\cos 240 = -0.5$ (where 240 is in degrees)

So, you can decide to work in radians or degrees.

13. The equation of the displacement of a body in simple harmonic motion is given by:
$$x = 2 \cos(5t + 1)$$
where x is in meters and t is in seconds. Calculate:
(a) the maximum speed
(b) the maximum acceleration
(c) the displacement of the body between time t = 0 and t = 1 second

Solution

(a) The given equation: $x = 2 \cos(5t + 1)$
The general equation: $x = A \cos(\omega t + \phi)$
Comparing the two equations above shows that the amplitude, A = 2m, while the angular velocity, $\omega = 5$
Hence, the maximum speed is given by:
$$v = \omega A$$
$$= 5 \times 2$$
$$v = 10 \text{ m/s}$$

(b) The maximum acceleration is given by:
$$a = \omega^2 A$$
$$= 5^2 \times 2$$
$$a = 50 \text{ ms}^{-2}$$

(c) The equation for the position of the body is given by:
$$x = 2 \cos(5t + 1)$$
At time t = 0, the position is obtained by substituting 0 for t in the equation above. This gives:
$$x = 2 \cos(5(0) + 1)$$
$$= 2 \cos 1$$
$$= 2 \times (0.5403) \quad \text{(Note that cos 1 = 0.5403, where 1 is in radians not degree)}$$
$$x = 1.08 \text{ m}$$
At time t = 1, the position is obtained by substituting 1 for t in the equation above. This gives:
$$x = 2 \cos(5(1) + 1)$$
$$= 2 \cos 6$$
$$= 2 \times (0.9602)$$
$$x = 1.92 \text{ m}$$
Therefore, between t = 0 to t = 1 sec, the body has been displace by:
$$1.92 - 1.08 = 0.84 \text{ m}$$

Exercise 1

1. A body undergoing simple harmonic motion has a maximum displacement from the center of motion as 10cm. If its angular velocity is 2rad/sec, calculate:
(a) the period
(b) the frequency
(c) the maximum velocity
(d) the acceleration at the center and at the end of motion
(e) the velocity of the body at a point 5cm from the center of motion
   (Take π = 3.142)

2. A simple pendulum of length 100cm oscillates with amplitude of 10cm. Calculate:
(a) the period of oscillation
(b) the maximum velocity of the motion   (Take g = 10m/s^2)

3. An object in simple harmonic motion has an angular velocity of 8rad/sec. If the amplitude of the motion is 15cm, calculate the velocity of the object at a point:
(a) 12cm from the equilibrium position
(b) 7cm from the end of motion

4. The acceleration of body in simple harmonic motion is 20 times its displacement in meters. Determine the frequency of the motion.

5. The period of a simple pendulum X is 5sec. What is the period of a simple pendulum Y which makes 50 vibrations in the time it takes X to make 60 vibrations.

6. An object moving with simple harmonic motion has an amplitude of 6cm and a frequency of 50Hz. Calculate:
(a) the period of oscillation
(b) the velocity at the middle and end of oscillation

7. The time it takes a simple pendulum to complete one vibration is 2.8sec. When the length of the pendulum is reduced by 2m, the period is 1.6sec. Determine:
(a) the original length of the pendulum
(b) the value of the acceleration due to gravity of the place.

8. The period of a simple pendulum is 20sec. Calculate its period when its length is doubled.

9. The period of a spiral spring is 8sec when the mass attached to it is 50g. What is its period when this mass is replaced with a 40g mass?

10. A simple pendulum of length 60cm performs simple harmonic motion. Determine the angular velocity of the pendulum. (g = 10m/s^2)

11. A spiral spring has a mass of 120g attached to it, and this extends the spring by 10cm. The spring is pulled down a distance of 6cm and allowed to perform simple harmonic motion. Calculate:
(a) the angular velocity of the motion

(b) the velocity of the spring at the point 2cm from the mean position.
   (g = 10m/s^2)

12. The equation of the displacement in meters, of a body performing simple harmonic motion is given by:

$$x = 10 \cos(2\pi t + \frac{\pi}{4})$$

where t is in seconds. Determine:
(a) the amplitude, frequency and period of motion
(b) the equations of the velocity and acceleration of the body
(c) position, velocity and acceleration of the body at time t = 2 seconds

13. The equation of the displacement of a body in simple harmonic motion is given by:

$$x = 8 \cos(4t + 3)$$

where x is in meters and t is in seconds. Calculate:
(a) the maximum speed
(b) the maximum acceleration
(c) the displacement of the body between time t = 0 and t = 0.5 seconds

14. The period of a simple pendulum is 12sec. Calculate its period when its length is quartered.

15. Two different simple pendulums perform simple harmonic motion. The ratio of their periods is 2 : 1. Find the ratio of the length of the shorter one to the longer one.

CHAPTER 2
ENERGY IN SIMPLE HARMONIC MOTION

If a spiral spring has a mass m attached to it, and it is pulled down a distance of A, and then released to perform simple harmonic motion, then the maximum potential energy of the motion is given by:

$$P.E_{max} = \frac{1}{2}kA^2$$

where A = amplitude, $k = \frac{F}{e}$, and it is the force constant of the spring, while e is the extension produced by the mass attached.

The potential energy of the motion at a point x from the center of motion is given by:

$$P.E = \frac{1}{2}kx^2$$

This simplifies to: $P.E = \frac{1}{2}m\omega^2 x^2$ (since $k = m\omega^2$)

The maximum kinetic energy of the motion is given by:

$$K.E_{max} = \frac{1}{2}mv^2$$

This simplifies to:

$$K.E_{max} = \frac{1}{2}m\omega^2 A^2 \quad \text{(since } v = \omega A\text{)}$$

However, the kinetic energy at a point x from the fixed point is given by:

$$K.E = \frac{1}{2}k(A^2 - x^2) \quad \text{or} \quad K.E = \frac{1}{2}m\omega^2(A^2 - x^2)$$

Potential energy is maximum at the ends of motion and zero at the center of motion while kinetic energy is maximum at the center of motion and zero at the ends of motion. The maximum kinetic energy is equal to the maximum potential energy.

Hence: $P.E_{max} = K.E_{max}$

Note that at any point in the motion of a body in SHM:

$$P.E + K.E = \text{Constant} = P.E_{max} \text{ and } K.E_{max}$$

Examples

1. A light spiral spring is loaded with a mass of 0.05kg and it extends by 0.1m. It is then pulled vertically down by 0.08m and allowed to perform simple harmonic motion. Calculate:
(a) the period of the spring
(b) the maximum potential energy of the motion
(c) the maximum kinetic energy of the motion
   (g = 10m/s^2)

Solution

(a) $T = 2\pi\sqrt{\dfrac{e}{g}}$

$= 2 \times 3.142 \times \sqrt{\dfrac{0.1}{10}}$

$= 6.284 \times \sqrt{0.01}$

$= 6.284 \times 0.1$

$T = 0.628$ sec.

(b) $P.E_{max} = \dfrac{1}{2}kA^2$

But, $k = \dfrac{F}{e}$

$= \dfrac{mg}{0.1}$ (Note that F = W = mg, where W is weight)

$= \dfrac{0.05 \times 10}{0.1}$

$k = 5 N/m$

Hence, $P.E_{max} = \dfrac{1}{2}kA^2$

$= \dfrac{1}{2} \times 5 \times 0.08^2$ (Note that the amplitude is 0.08m)

$= \dfrac{1 \times 5 \times 0.0064}{2}$

$P.E = 0.016 J$

(c) $K.E_{max} = \dfrac{1}{2}m\omega^2 A^2$

But, $\omega = \dfrac{2\pi}{T}$

$= \dfrac{2 \times 3.142}{0.628}$

$\omega = 10$ rad/sec

Hence, $K.E_{max} = \dfrac{1}{2}m\omega^2 A^2$

$= \dfrac{1}{2} \times 0.05 \times 10^2 \times 0.08^2$

$= \dfrac{1 \times 0.05 \times 100 \times 0.0064}{2}$

$K.E_{max} = 0.016 J$

This shows that the maximum kinetic energy is equal to the maximum potential energy.

2. When a load of 20g is attached to a spiral spring, an extension of 0.04m is produced. If

the string is pulled vertically down by 18cm and allowed to perform SHM, calculate:
(a) the period of oscillation
(b) the kinetic energy of the motion at 14cm from the center of oscillation.

Solution

(a) Given: m = $\frac{20}{1000}$ = 0.02kg, A = $\frac{18}{100}$ = 0.18m, e = 0.04m.

$T = 2\pi \sqrt{\frac{e}{g}}$

$= 2 \times 3.142 \times \sqrt{\frac{0.04}{10}}$

$= 6.284 \times \sqrt{0.004}$

$= 6.284 \times 0.0632$

T = 0.397sec.

(b) K.E = $\frac{1}{2}k(A^2 - x^2)$

But, k = $\frac{F}{e}$

$= \frac{mg}{0.1}$

$= \frac{0.02 \times 10}{0.04}$

k = 5N/m

Hence, K.E = $\frac{1}{2}k(A^2 - x^2)$

$= \frac{1}{2} \times 5(0.18^2 - 0.14^2)$ (Note that x = 14cm = 0.14m)

= 2.5(0.0324 – 0.0196)

K.E = 0.032J

3. A particle of mass 10g performs simple harmonic motion of amplitude 6cm and period 2π seconds. Calculate:
(a) the kinetic energy and potential energy when it is at a distance of 4cm from its equilibrium position.
(b) the maximum kinetic energy of the particle.

Solution

Given: m = $\frac{10}{1000}$ = 0.01kg, A = $\frac{6}{100}$ = 0.06m, x = 0.04m.

(a) K.E = $\frac{1}{2}m\omega^2(A^2 - x^2)$

But, $\omega = \dfrac{2\pi}{T}$

$= \dfrac{2\pi}{2\pi}$

$\omega = 1\,\text{rad/sec}$

Hence, $K.E = \dfrac{1}{2}m\omega^2(A^2 - x^2)$

$= \dfrac{1}{2} \times 0.01 \times 1^2 (0.06^2 - 0.04^2)$

$= \dfrac{1 \times 0.01 \times 1 \times 0.002}{2}$

$= 0.00001\,J$

The potential energy is given by:

$P.E = \dfrac{1}{2}m\omega^2 x^2$

$= \dfrac{1}{2} \times 0.01 \times 1^2 \times 0.04^2$

$= \dfrac{1 \times 0.01 \times 1 \times 0.0016}{2}$

$= 0.000008\,J$

(b) Recall that at any point:

P.E + K.E = Constant = $P.E_{max}$ and $K.E_{max}$

Hence, P.E + K.E = $K.E_{max}$

$0.00001 + 0.000008 = K.E_{max}$

$K.E_{max.} = 0.000018\,J$

4. A body of mass 20g performs simple harmonic motion at a frequency of 5Hz. At a distance of 10cm from the mean position, its velocity is 200cm/s. Calculate its:
(a) maximum displacement from the mean position
(b) potential and kinetic energy at this position of 10cm from the mean position
(c) maximum potential and kinetic energy.
   ($\pi = 3.14$, $g = 10\,m/s^2$)

Solution

Given: $m = \dfrac{20}{1000} = 0.02\,kg$, $f = 5\,Hz$, $x = 0.1\,m$, $v = \dfrac{200}{100} = 2\,m/s$

(a) Let us first determine the angular velocity as follows:

$\omega = 2\pi f$

$= 2 \times 3.14 \times 5$

$\omega = 31.4\,\text{rad/sec}$

Recall that: $v = \omega\sqrt{A^2 - x^2}$

$2 = 31.4\sqrt{A^2 - 0.1^2}$

Squaring both sides of the equation gives:

$2^2 = 31.4^2(A^2 - 0.1^2)$

$4 = 985.96(A^2 - 0.01)$

$4 = 985.96A^2 - 9.8596$

$4 + 9.8596 = 985.96A^2$

$A = \sqrt{\dfrac{13.8596}{985.96}}$

$A = 0.119$

Therefore, the maximum displacement from the mean position is 0.119m

(b) $P.E = \dfrac{1}{2}m\omega^2 x^2$

$= \dfrac{1}{2} \times 0.02 \times 31.4^2 \times 0.1^2$

$= \dfrac{1 \times 0.02 \times 985.96 \times 0.01}{2}$

$P.E = 0.0986J$

$K.E = \dfrac{1}{2}m\omega^2(A^2 - x^2)$

$= \dfrac{1}{2} \times 0.02 \times 31.4^2(0.119^2 - 0.1^2)$

$= \dfrac{1 \times 0.02 \times 985.96 \times 0.004161}{2}$

$K.E = 0.0821J$

(c) Recall that at any point:

$P.E + K.E = $ Constant $= P.E_{max}$ and $K.E_{max}$

Hence, $P.E + K.E = P.E_{max}$ and $K.E_{max}$

$0.0986 + 0.0821 = P.E_{max}$ and $K.E_{max}$

$P.E_{max}$ and $K.E_{max.} = 0.181J$

Therefore, the maximum potential and kinetic energy are 0.181J each.

5. The mass of the bob of a simple pendulum performing simple harmonic motion is 15g. If at the ends of motion, the bob is at a vertical distance of 2cm from the equilibrium point, calculate the maximum kinetic energy of the bob. ($g = 10m/s^2$)

Solution

Given: $m = 0.015kg$, $h = 0.02m$ and $g = 10m/s^2$

The maximum potential energy of the bob is obtained by:

$P.E_{max} = mgh$

= 0.015 x 10 x 0.02
= 0.003J

Recall that: $P.E_{max} = K.E_{max}$

Therefore, the maximum kinetic energy of the bob is 0.003J.

6. A 5kg mass is suspended from the end of a spring and released to perform simple harmonic motion with amplitude 4cm and a period of 2sec. Find the maximum energy of the mass.

<u>Solution</u>

Let us determine the angular velocity/frequency as follows:

$\omega = \dfrac{2\pi}{T}$

$= \dfrac{2 \times 3.142}{2}$

$\omega = 3.142 \text{rads}^{-1}$

Hence the energy of the mass is given by:

$E = \dfrac{1}{2}m\omega^2 A^2$

$= \dfrac{1}{2} \times 5 \times 3.142^2 \times 0.04^2$ (Note that 4cm = 0.04m)

$= \dfrac{1 \times 5 \times 9.872 \times 0.0016}{2}$

$E = 0.0395J$

Exercise 2

1. A mass of 80g is hung from a spiral spring thereby producing an extension of 0.02m. It is then pulled vertically down by 0.05m and allowed to perform simple harmonic motion. Calculate:
(a) the period of the spring
(b) the maximum potential energy of the motion
(c) the maximum kinetic energy of the motion
   (g = 10m/s^2)

2. When a load of 30g is attached to a spiral spring, an extension of 4cm is produced. If the string is pulled vertically down by 8cm and allowed to perform SHM, calculate:
(a) the period of oscillation
(b) the kinetic energy of the motion at 2cm from the center of oscillation.

3. A particle of mass 2g performs simple harmonic motion of amplitude 3cm and period π seconds. Calculate:
(a) the kinetic energy and potential energy when it is at a distance of 1cm from its

equilibrium position.

(b) the maximum potential energy of the particle.

4. A body of mass 10g performs simple harmonic motion of frequency of 2Hz. At a distance of 5cm from the mean position, its velocity is 0.8m/s. Calculate its:

(a) maximum displacement from the mean position

(b) potential and kinetic energy at this position of 5cm from the mean position

(c) maximum potential and kinetic energy.

$(g = 10m/s^2)$

5. The mass of the bob of a simple pendulum performing simple harmonic motion is 9g. If at the ends of motion, the bob is at a vertical distance of 2.4cm from the equilibrium point, calculate the maximum kinetic energy of the bob. $(g = 10m/s^2)$

6. A 200g mass is suspended from the end of a spring and released to perform simple harmonic motion with amplitude 12cm and a period of 16sec. Find the maximum energy of the mass.

7. A particle of mass 6g performs simple harmonic motion of amplitude 5cm and period 10 seconds. Calculate:

(a) the kinetic energy and potential energy when it is at a distance of 2cm from its equilibrium position.

(b) the maximum potential energy of the particle.

8. A spring is loaded with a mass of 0.02kg and it extends by 0.08m. It is then pulled vertically down by 4cm and allowed to perform simple harmonic motion. Calculate:

(a) the period of the spring

(b) the maximum potential energy of the motion

(c) the maximum kinetic energy of the motion

$(g = 10m/s^2)$

CHAPTER 3
CIRCULAR MOTION

When a body moves round a circular path, its motion is described as a circular motion. Its path is a curve which gives an angular distance, θ, measured in radians. Hence the angular velocity, ω, of the body is defined as follows:

$$\omega = \frac{\text{Angle turned through by the body}}{\text{time taken}}$$

$$\omega = \frac{\theta}{t}$$

where θ = angular distance in radians, and t = time in seconds.
The circular distance, s, or length of arc that the body has moved can be obtained by:

$$s = r\theta$$

where r is the radius of the circular path.
Note that 360 degrees which is one cycle is equal to 2π radians or $180° = \pi$ radians. Hence, an angle, α, in degrees can be converted to angle in radians by using the formula below:

$$\theta = \frac{\alpha \pi}{180}$$ (Note that the value of π is usually taken to be 3.142)

The relationship between linear velocity and angular velocity for a body undergoing circular motion is given by:

$$\omega = \frac{v}{r}$$

Or, $v = \omega r$

This shows that circular motion is related to simple harmonic motion (where $v = \omega A$). Generally, the formulas for simple harmonic motion can be used for circular motion. Hence in circular motion, the linear acceleration of the body is given by:

$$a = \omega^2 r$$

Or, $a = \dfrac{v^2}{r}$

Linear acceleration is related to angular acceleration by the expression below:

$$\alpha = \frac{a}{r}$$

Or, $a = \alpha r$

where α is the angular acceleration measured in per square second (s^{-2})

Centripetal Force

When a body moves round a circular path, it experiences a force which is directed towards the center of the circle and keeps the body on the circular path. This force is called centripetal force. It is given by:

$$F = m\omega^2 r \quad (\text{since } F = ma \text{ and } a = \omega^2 r)$$

Or, $F = \dfrac{mv^2}{r}$ (since a is also $\dfrac{v^2}{r}$)

Examples

1. A body moves along a circular path with a uniform angular speed of 0.5rad/sec and at a constant speed of 1m/s. Calculate the acceleration of the body.

Solution

Let us first calculate the radius of the path.

$v = \omega r$

$r = \dfrac{v}{\omega}$

$= \dfrac{1}{0.5}$

$r = 2m$

Therefore, the acceleration is given by:

$a = \omega^2 r$

$= 0.5^2 \times 2$

$= 0.25 \times 2$

$a = 0.5 ms^{-2}$

2. A body rotating on a circular path makes 420r.p.m. What is its frequency?

Solution

420r.p.m means 420 revolutions per minute. This means that the body made 420 cycles (i.e. revolutions) in 60 seconds (i.e. 1 minute).

Recall that: $f = \dfrac{\text{Number of oscillations /cycles}}{\text{time taken}}$

$f = \dfrac{420}{60}$

$f = 6Hz$

3. An object covers an angular distance of 270° in 8 seconds. Calculate its angular velocity in rad/sec.

Solution

Let us first convert the angle from degrees to radians as follows:

$\theta = \dfrac{\alpha \pi}{180}$ (Note that $\alpha = 270°$)

$= \dfrac{270 \times 3.142}{180}$

$\theta = 4.713$ radians

Therefore, the angular velocity is obtained as follows:

$$\omega = \frac{\theta}{t}$$
$$= \frac{4.713}{8}$$
$$= 0.589 \text{ radsec}^{-1}$$

4. A particle makes 300rpm on a circular path of radius 60cm. Find:
(a) its period
(b) its angular velocity
(c) its linear velocity
(d) its acceleration
(e) its angular acceleration

Solution

(a) 300rpm means 300 revolutions per minute. This means that the body made 300 cycles (i.e. revolutions) in 60 seconds (i.e. 1 minute).

Recall that: $T = \frac{\text{time taken}}{\text{Number of cycles}}$

$$T = \frac{60}{300}$$
$$T = 0.2 \text{sec}$$

(b) $\omega = \frac{2\pi}{T}$
$$= \frac{2 \times 3.142}{0.2}$$
$$\omega = 31.42 \text{ radsec}^{-1}$$

(c) $v = \omega r$
$$= 31.42 \times 0.6 \quad (60\text{cm} = 0.6\text{m})$$
$$= 18.852 \text{ ms}^{-1}$$

(d) $a = \omega^2 r$
$$= 31.42^2 \times 0.6$$
$$= 987.2164 \times 0.6$$
$$= 592.3 \text{ ms}^{-2}$$

(e) Angular acceleration is given by:
$$\alpha = \frac{a}{r}$$
$$= \frac{592.3}{0.6}$$

$= 987.2s^{-2}$

5. An object in a circular path covers an angle of 315° in 6sec. If the radius of the circular path is 10cm, calculate:
(a) its angular speed in rad/sec
(b) its linear velocity
(c) its frequency
(d) its maximum acceleration

Solution
(a) Let us first convert the angle from degrees to radians as follows:

$$\theta = \frac{\alpha\pi}{180}$$

$$= \frac{315 \times 3.142}{180}$$

$\theta = 5.4985$ radians

Therefore, the angular speed is obtained as follows:

$$\omega = \frac{\theta}{t}$$

$$= \frac{5.4985}{6}$$

$= 0.916$ radsec^{-1}

(b) $v = \omega r$
$= 0.916 \times 0.1$ (10cm = 0.1m)
$= 0.0916$ ms^{-1}

(c) $f = \frac{\omega}{2\pi}$

$= \frac{0.916}{2 \times 3.142}$

$= 0.146$ Hz

(d) $a = \omega^2 r$
$= 0.916^2 \times 0.1$
$= 0.0839$ ms^{-2}

6. A body of mass 50g moves a velocity of 2m/s in a circular path of radius 20cm. Calculate the centripetal force acting on the body.

Solution
Given: m = 0.05kg, v = 2m/s, r = 0.2m
The centripetal force on the body is given by:

$$F = \frac{mv^2}{r}$$
$$= \frac{0.05 \times 2^2}{0.2}$$
$$F = 1N$$

7. A particle of mass 4g makes 540rpm on a circular path of radius 6cm. Calculate the value of the force that keeps the body on the path.

Solution

Recall that: $f = \frac{\text{Number of cycles}}{\text{time taken}}$

$$= \frac{540}{60}$$
$$f = 9Hz$$
$$\omega = 2\pi f$$
$$= 2 \times 3.142 \times 9$$
$$\omega = 56.556 \text{ radsec}^{-1}$$

The force required is given by:
$$F = m\omega^2 r$$
$$= 0.004 \times 56.556^2 \times 0.06$$
$$F = 0.768N$$

8. An object of mass 0.2kg covers an angular distance of 90° in 1.2 seconds. If its path is a circle of radius 0.4m, calculate:
(a) its angular frequency
(b) its linear velocity
(c) the centripetal force acting on the object.

Solution

(a) Let us first convert the angle from degrees to radians as follows:
$$\theta = \frac{\alpha\pi}{180}$$
$$= \frac{90 \times 3.142}{180}$$
$$\theta = 1.571 \text{ radians}$$

Therefore, the angular frequency is obtained as follows:
$$\omega = \frac{\theta}{t}$$
$$= \frac{1.571}{1.2}$$
$$\omega = 1.309 \text{ radsec}^{-1}$$

(b) v = ωr
= 1.309 x 0.4
= 0.524ms^{-1}

(c) The centripetal force is given by:
F = mω²r
= 0.2 x 1.309² x 0.4
F = 0.137N

Or, the centripetal force can also be given by:

$$F = \frac{mv^2}{r}$$

$$= \frac{0.2 \times 0.524^2}{0.4}$$

F = 0.137N   (As obtained above)

Exercise 3

1. A body moves along a circular path with a uniform velocity of 1rad/sec and at a constant speed of 2m/s. Calculate the acceleration of the body.
2. A body rotating on a circular path makes 2400r.p.m. What is the frequency of the body?
3. An object covers an angular distance of 60° in 5 seconds. Calculate its angular velocity in rad/sec.
4. A particle makes 360rpm on a circular path of radius 20cm. Find:
(a) its period
(b) its angular velocity
(c) its linear velocity
(d) its acceleration
(e) its angular acceleration
5. An object in a circular path covers an angle of 180° in 4sec. If the radius of the circular path is 80cm, calculate:
(a) its angular speed in rad/sec
(b) its linear velocity
(c) its frequency
(d) its maximum acceleration
6. A body of mass 110g moves at a velocity of 300cm/s in a circular path of radius 8cm. Calculate the centripetal force acting on the body.
7. A particle of mass 14g makes 200rpm on a circular path of radius 2cm. Calculate the value of the force that keeps the body on the path.

8. An object of mass 10g covers an angular distance of 100° in 1.8 seconds. If its path is a circle of radius 0.12m, calculate:

(a) its angular frequency

(b) its linear velocity

(c) the centripetal force acting on the object.

CHAPTER 4
WAVE MOTION

Wave is a disturbance that travels through a medium and transfers energy from one point to another without actually causing a permanent displacement of the medium.

The period of a wave is given by:

$$T = \frac{2\pi}{w}$$ where w = angular velocity. $w = 2\pi f$. This simplifies to also give:

$$T = \frac{1}{f}$$ where f is the frequency of the wave.

The wave velocity of a wave is given by:

$$v = \lambda/T$$ this simplifies to:

$$v = f\lambda \quad (\text{Since } T = \frac{1}{f})$$

This equation is applicable to all waves.

A wave can also be represented mathematically as follows:

$$y = A \sin\phi \quad \text{where } \phi = \text{phase angle}$$

But $\phi = \omega(t - \frac{x}{v})$ where ω = angular velocity, t = time, x = horizontal distance, v = velocity and y = vertical distance

$$\therefore \quad y = A \sin \omega(t - \frac{x}{v})$$

Since $\omega = 2\pi f$, the equation above can be expressed as follows:

$$y = A \sin 2\pi f (t - \frac{x}{v})$$

Or $\quad y = A \sin (2\pi ft - \frac{2\pi fx}{v})$

Or $\quad y = A \sin (2\pi ft - 2\pi x/\lambda) \quad (\text{Since } \frac{f}{v} = 1/\lambda \text{ from } v = f\lambda)$

The expression $2\pi/\lambda$ in the equation above is called the wave number, k.

The two wave equations that will be used in the solved examples below are:

$$y = A \sin (2\pi ft - \frac{2\pi fx}{v}) \quad \text{and}$$

$$y = A \sin (2\pi ft - 2\pi x/\lambda)$$

Examples

1. On a graphical representation of a wave the height of each curve above the x axis is 8m, while the width of each curve along the x axis is 5m.
   (a) Find the amplitude and wavelength of the motion
   (b) Calculate the period of the wave if it has an angular velocity of 6.4rad/sec
   (c) Calculate the velocity of the wave

Solution

(a) The amplitude is the maximum vertical height of the wave. So, amplitude A, is 8m

The wavelength is the width of two curves. This width is also equal to the distance between successive wave crest. So, wavelength, λ = 2 x 5 = 10m

(b) w = 6.4rad/sec. But period, T = $\frac{2\pi}{w}$

∴ T = $\frac{2\pi}{6.4}$ = $\frac{2 \times 3.142}{6.4}$

= $\frac{6.284}{6.4}$ = 0.982

The period is 0.982sec.

(c) v = λ/T

= $\frac{10}{0.982}$ = 10.2

The velocity of the wave is 10.2ms^{-1}

2. Determine the wavelength of a wave motion that travels a distance of 6m which is made up of five curves.

Solution

The width of each of the bell shape curve of a wave is equal to half a wavelength.

∴ Since 1 curve = λ/2, then 5 curves = 5 x λ/2 = 5λ/2

These 5 curves are 6m wide. This means that:

5λ/2 = 6

∴ 5λ = 6 x 2

λ = $\frac{12}{5}$ = 2.4

The wavelength of the wave is 2.4m.

3. A wave having a frequency of 25Hz travels at a velocity of 40m/s. Calculate the wavelength of the motion.

Solution

v = fλ

∴ λ = $\frac{v}{f}$

= $\frac{40}{25}$ = 1.6

The wavelength is 1.6m

4. A wave travels a distance of 42m in 7sec, and the distance between successive wave crests of the wave is 1.2m. Calculate the frequency of the wave.

Solution

$$\text{Velocity} = \frac{Distance}{Time}$$

$$\therefore \quad v = \frac{42}{7} = 6m/s$$

Also, $v = f\lambda$

So, $f = v/\lambda$

$$= \frac{6}{1.2} = 5$$

The frequency of the wave is 5Hz.

5. The distance between successive troughs of a wave of frequency 20Hz is 40cm. Calculate the time taken by the wave to cover a distance of 26m.

Solution

Wavelength, λ, is the distance between successive troughs.

$$\therefore \quad \lambda = 40cm = (\frac{40}{100})m = 0.4m, \text{ and } f = 20Hz$$

So, $v = f\lambda$

$$= 20 \times 0.4 = 8ms^{-1}$$

But, $v = \frac{Distance}{Time}$

$$\therefore \quad 8 = \frac{26}{Time}$$

$$Time = \frac{26}{8} = 3.25$$

The time taken by the wave is 3.25sec.

6. The displacement, y, of a wave travelling in the negative x direction is represented by, $y = 5\sin 12\pi (t + \frac{x}{30})$, where x and y are in metres and t in seconds.

(a) What is the amplitude of the wave?
(b) Determine the frequency of the wave
(c) Determine the velocity of the wave
(d) What is the wavelength of the wave?
(e) Find the wave number of the wave.

Solutions

(a) The general equation of a wave is: $y = A \sin (2\pi ft - \frac{2\pi fx}{v})$

The equation of the wave in the question is: $y = 5\sin 12\pi (t + \frac{x}{30})$

Comparing the two equations shows that the amplitude A = 5

∴ The amplitude of the wave is 5m.

(b) $y = 5\sin 12\pi (t + \frac{x}{30})$. When 12π is used to expand the bracket, the equation becomes:

$y = 5\sin (12\pi t + \frac{12\pi x}{30})$. The general wave equation is: $y = A \sin (2\pi ft - \frac{2\pi fx}{v})$

Comparing these two equations shows that the first terms in both brackets which contain the frequency, f, (i.e. what we want to find) and the time, t, are equal. Note that only the general wave equation will contain what we want to calculate, i.e. the frequency, f. Both terms contain the time, t. Equating them gives:

$2\pi ft = 12\pi t$. Cancelling out the t and π gives:

$2f = 12$

∴ $f = \frac{12}{2} = 6$

The frequency of the wave is 6Hz.

(c) The expanded equation from the question and the general wave equation are:

$y = 5\sin (12\pi t + \frac{12\pi x}{30})$ and, $y = A \sin (2\pi ft - \frac{2\pi fx}{v})$

In order to determine the velocity, compare the two equations and equate the equal terms based on what we want to calculate. The terms are those that contain x, and the velocity, v, (i.e. what we want to find).

So, $\frac{12\pi x}{30} = \frac{2\pi fx}{v}$. Cancelling out the x and π gives:

$\frac{12}{30} = \frac{2f}{v}$

$12v = 30 \times 2f$

$12v = 60f$

$v = \frac{60f}{12}$

$= \frac{60 \times 6}{12}$ (Since f = 6 from solution (b) above)

$= \frac{360}{12} = 30$

So, the velocity of the wave is 30ms^{-1}

(d) The general wave equation that contains wavelength is: $y = A \sin (2\pi ft - 2\pi x/\lambda)$

Also, from the question, $y = 5\sin (12\pi t + \frac{12\pi x}{30})$

In order to determine the wavelength equate the terms that contain the wavelength, λ, along with the common distance x.

So, $2\pi x/\lambda = \frac{12\pi x}{30}$. Cancelling out the x and π gives:

$$\frac{2}{\lambda} = \frac{12}{30}$$
$$12\lambda = 2 \times 30$$
$$\lambda = \frac{60}{12} = 5$$

The wavelength is 5m.

(e) The wave number, k, is given by:
$$k = \frac{2\pi}{\lambda}$$
$$= \frac{2 \times 3.142}{5} = \frac{6.284}{5} = 1.26$$

The wave number is $1.26 m^{-1}$

7. The frequency of a travelling wave is 10Hz. Starting from the origin, its first crest is at $x = 2m$.
(a) What is the wavelength of the wave?
(b) What time does it take the wave to attain its first crest?
(c) Determine the velocity of the wave
(d) If the amplitude of the wave is 5m, what is the equation of the wave?

Solution

(a) The first crest is half of a semi circle, which is also a quarter of a circle. A cycle which gives a wavelength is equal to four quarters of a circle. From the question above, a quarter of a circle (i.e. first crest) is 2m long. Hence a wavelength which is four quarters is given by:
$$\lambda = 4 \times 2$$
$$\lambda = 8m$$

The wavelength is 8m

(b) The time taken to travel one wavelength is equal to the period, and the period is given by: $T = \frac{1}{f}$
$$= \frac{1}{10}$$
$$T = 0.1 sec$$

∴ It takes 0.1sec to travel a distance of λ, i.e. 8m.

But, the first crest is attained at a quarter of one wavelength, i.e. $\lambda/4$

Similarly, it will also take a quarter of a period to attain the first crest. This is given by:
$$\text{Time taken} = \frac{T}{4}$$
$$= \frac{0.1}{4}$$
$$= 0.025$$

Hence, it takes 0.025sec for the wave to attain its first crest.

(c) The velocity of the wave is given by:
$$v = f\lambda$$
$$= 30 \times 8 = 240$$
The velocity of the wave is 240m/s

(d) The general equation of a wave is given by:
$$y = A \sin(2\pi ft - 2\pi x/\lambda)$$
Factorizing the term in the bracket gives:
$$y = A \sin 2\pi (ft - x/\lambda)$$
Substituting A = 5, f = 10 and λ = 8, gives:
$$\therefore \quad y = 5\sin 2\pi \left(10t - \frac{x}{8}\right)$$
It can also be expressed by substituting the actual value of π as follows:
$$y = 5\sin 2\pi \left(10t - \frac{x}{3}\right)$$
$$y = 5\sin 2 \times 3.142 \left(10t - \frac{x}{3}\right)$$
$$y = 5\sin 6.284 \left(10t - \frac{x}{3}\right)$$
$$y = 5\sin \left(6.284 \times 10t - \frac{6.284x}{3}\right)$$
$$\therefore \quad y = 5\sin(62.8t - 2.1x)$$

8. If y = 2sin (3x – 4t), where x and y are in metres and t in seconds represent a wave motion, Determine the:
(a) amplitude
(b) frequency
(c) period
(d) velocity
(e) angular velocity
(f) wavelength
(g) wave number of the wave.

Solutions

(a) The general equation of a wave is: $y = A \sin\left(2\pi ft - \frac{2\pi fx}{v}\right)$

The equation of the wave in the question is: y = 2sin (3x – 4t)
Comparing the two equations shows that the amplitude A = 2
∴ The amplitude of the wave is 2m.

(b) y = 2sin (3x – 4t). The general wave equation is: y = A sin (2πft - $\frac{2\pi fx}{v}$)

Comparing these two equations shows that we equate the terms in both brackets which contain the time t and the frequency, f, (i.e. what we want to find). Note that only the general wave equation will contain what we want to calculate, i.e. the frequency, f. Equating the terms gives:

2πft = 4t.   (Note that the negative sign should be ignored).

Cancelling out the t gives:

2πf = 4

∴   f = $\frac{4}{2\pi}$ = $\frac{4}{2 \times 3.142}$ = $\frac{4}{6.284}$ = 0.64

The frequency of the wave is 0.64Hz.

(c) The period is given by:

T = $\frac{1}{f}$

= $\frac{1}{0.64}$ = 1.56

The period is 1.56sec.

(d) y = 2sin (3x – 4t) and y = A sin (2πft - $\frac{2\pi fx}{v}$)

In order to determine the velocity, compare the two equations and equate the equal terms based on what we want to calculate. The terms are those that contain x in the two brackets, and the velocity, v, (i.e. what we want to find) in only one bracket.

So, 3x = $\frac{2\pi fx}{v}$.   Cancelling out the x gives:

3 = $\frac{2\pi f}{v}$

3v = 2πf

∴   v = $\frac{2\pi f}{3}$

= $\frac{2 \times 3.142 \times 0.64}{3}$ (Since f = 0.64 from solution (b) above)

= $\frac{4.022}{3}$ = 1.34

The velocity of the wave is 1.34ms^{-1}

(e) The angular velocity, w, is given by:

ω = 2πf

= 2 x 3.142 x 0.64

= 4.02rad/sec

(f) The general wave equation that contains wavelength is: $y = A \sin(2\pi ft - 2\pi x/\lambda)$

Also, from the question, $y = 2\sin(3x - 4t)$

In order to determine the wavelength equate the terms that contain the wavelength, λ, (i.e. what we want to find) along with the common distance x.

So, $2\pi x/\lambda = 3x$. Cancelling out the x gives:

$2\pi/\lambda = 3$

$3\lambda = 2\pi$

$\lambda = \dfrac{2\pi}{3} = \dfrac{2 \times 3.142}{3} = \dfrac{6.284}{3} = 2.09$

The wavelength is 2.09m.

(g) The wave number, k, can also be obtained by comparing the wave equations as follows:

$y = 2\sin(3x - 4t)$ and $y = A\sin(2\pi ft - 2\pi x/\lambda)$

Since $k = 2\pi/\lambda$, then by comparison,

$2\pi x/\lambda = 3x$

∴ $kx = 3x$ (By substituting $2\pi/\lambda$ for k)

Cancelling out x shows that:

$k = 3$

The wave number is 3m^{-1}

9. Waves whose crests are 40cm apart made a cork floating on water in a ripple tank to rise and fall through a total range of 5cm once every 2 seconds. Determine the:
(a) amplitude
(b) frequency
(c) velocity of the wave.

Solutions

(a) The range of 5cm is the vertical distance between the crest and the trough. It is twice the amplitude.

∴ Amplitude = $\dfrac{1}{2}$(range) = $\dfrac{1}{2} \times 5$ = 2.5cm

(b) The period is the time taken to rise from the starting horizontal level up to the crest and then to fall to the trough, and finally to the starting level. This took a total time of 2 seconds. It is also the time taken to cover a distance of one wavelength. So, the period, T, is 2 seconds.

But, frequency, f is given by:

$f = \dfrac{1}{T}$

$= \dfrac{1}{2} = 0.5$

The frequency is 0.5Hz

(c) The velocity is given by:
   $v = f\lambda$
   But, $\lambda = 40cm = (\frac{40}{100})m = 0.4m$
   $\therefore \quad v = f\lambda$
   $= 0.5 \times 0.4 = 0.2$
   The velocity of the wave is $0.2ms^{-1}$

Exercise 4

1. On a graphical representation of a wave the height of each curve above the x axis is 200cm, while the width of each curve along the x axis is 80cm.
   (a) Find the amplitude and wavelength of the motion
   (b) Calculate the period of the wave if it has an angular velocity of 5rad/sec
   (c) Calculate the velocity of the wave

2. Determine the wavelength of a wave motion that travels a distance of 16m along the x-axis which is made up of ten curves.

3. A wave having a frequency of 40Hz travels at a velocity of 15m/s. Calculate the wavelength of the motion.

4. A wave travels a distance of 300cm in 3.25sec, and the distance between successive wave crests of the wave is 2m. Calculate the frequency of the wave.

5. The distance between successive troughs of a wave of frequency 50Hz is 6m. Calculate the time taken by the wave to cover a distance of 16m.

6. The displacement, y, of a wave travelling in the negative x direction is represented by, $y = 10\sin8\pi (t + \frac{x}{50})$, where x and y are in metres and t in seconds.
   (a) What is the amplitude of the wave?
   (b) Determine the frequency of the wave
   (c) Determine the velocity of the wave
   (d) What is the wavelength of the wave?
   (e) Find the wave number of the wave.

7. The frequency of a travelling wave is 25Hz. Its first crest is at $x = 5m$.
   (a) What is the wavelength of the wave?
   (b) What time does it take the wave to attain its first crest?
   (c) Determine the velocity of the wave
   (d) If the amplitude of the wave is 8m, what is the equation of the wave?

8. If y = 4sin (5x – 2t), where x and y are in metres and t in seconds represent a wave motion, Determine the:
(a) amplitude
(b) frequency
(c) period
(d) velocity
(e) angular velocity
(f) wavelength
(g) wave number of the wave.

9. Waves whose crests are 1.2m apart made a cork floating on water in a ripple tank to rise and fall through a total range of 0.14m once every 5 seconds. Determine the:
(a) amplitude
(b) frequency
(c) velocity of the wave.

10. The distance between five troughs of a wave is 4m. If the period of the wave is 0.8sec, find:
(a) the wavelength of the wave
(b) the velocity of the wave
(c) the equation of the wave if its amplitude is 1.5m

CHAPTER 5
ECHOES

Echo is a sound heard after the reflection of sound wave from a plane surface. Echo can be used to determine the speed of sound in air by using the expression:

$$v = \frac{2x}{t},$$

where v is the speed of sound in air, x is the distance between the source of sound and the reflecting surface, while t is the time in seconds taken to hear the echo.

The speed of sound in air is also proportional to the square root of the air's absolute temperature. This shows that:

$$v \propto \sqrt{T}$$

Examples

1. A man stands in front of a cliff and fires a gun. He hears the echo from the cliff after 2.5sec. If the speed of sound in air is 340m/s, how far is the man from the cliff?

<u>Solution</u>

$v = \frac{2x}{t}$ (Where x is the man's distance from the cliff)

$340 = \frac{2x}{2.5}$

∴ $2x = 340 \times 2.5$

$2x = 850$

$x = \frac{850}{2}$

$x = 425m$

The man is at a distance of 425m from the cliff

2. A horn is sounded from a wall, 1320m way. How long will it take to hear the reflected sound? (Speed of sound in air is 330m)

<u>Solution</u>

$v = \frac{2x}{t}$

$330 = \frac{2 \times 1320}{t}$

$330t = 2640$

∴ $t = \frac{2640}{330}$

$t = 8.5sec.$

It will take 8.5sec to hear the reflected sound

3. A whistle is blown at a distance of 252m from a vertical wall. If the echo of the sound produced by the whistle is heard after 1.5sec, calculate the speed of sound in air.
Solution
$$v = \frac{2x}{t}$$
$$= \frac{2 \times 252}{1.5}$$
$$= \frac{504}{1.5}$$
$$\therefore v = 336 ms^{-1}$$
The speed of sound in air is 336ms^{-1}

4. Two people Jane and Michael are 170m apart along a horizontal ground. A vertical wall is 510m behind Michael. Jane fires a gun. What is the time interval between the two sounds:
(a) heard by Michael
(b) heard by Jane.
(Speed of sound in air is 340m)
Solutions
(a) By the time Michael hears the reflected sound, the sound has travelled from Jane to Michael (170m), from Michael to the wall (510m) and from the wall to Michael (510m). This gives a total distance of:
170 + 510 + 510 = 1190
$$Velocity = \frac{Distance}{time}$$
$$340 = \frac{1190}{t}$$
340t = 1190
$$\therefore t = \frac{1190}{340}$$
t = 3.5sec

The time interval between the two sounds (first sound and echo) heard by Michael is 3.5sec.

(b) The distance of Jane from the wall is 170 + 510 = 680m
$$v = \frac{2x}{t}$$ (Where x is Jane's distance from the wall)
$$340 = \frac{2 \times 680}{t}$$
340t = 1360

∴ $t = \frac{1360}{340} = 4$

The time interval between the two sounds heard by Jane is 4sec.

5. A horn is sounded at regular intervals in front of a vertical wall 510m away. If the echo from the wall is heard simultaneously with the next hoot, how many hoots are made every minute. (Velocity of sound in air = 340ms^{-1})

Solution

$v = \frac{2x}{t}$

$340 = \frac{2 \times 510}{t}$

$340t = 2 \times 510$

$t = \frac{1020}{340} = 3$

The echo is heard every 3 seconds.

This is also the time taken to make each hoot. So, by simple proportion the number of hoots made in one minute is given by:

$\frac{60}{3} = 20$

The number of hoots made every minute is 20 hoots.

6. A boy 41m from a tall building claps his hands once every half second. He hears the echo of each clap midway between the clap and the next clap. Calculate the speed of sound.

Solution

The boy claps his hands every half second, i.e. $\frac{1}{2}$ second. He hears the echo midway between this time, i.e. $\frac{1}{2} \times (\frac{1}{2}$ second$) = \frac{1}{2} \times \frac{1}{2} = \frac{1}{4}$ seconds

So, it takes him $\frac{1}{4}$ seconds to hear the echo.

But, $v = \frac{2x}{t}$

$= \frac{2 \times 41}{\frac{1}{4}} = \frac{82}{0.25} = 328$

The speed of sound is 328ms^{-1}

7. The velocity of sound in air at a temperature of 25°C is 332ms^{-1}. What is the velocity when the temperature is 37°C?

Solution

In order to convert Celsius to Kelvin, add 273.

$v \, \alpha \, \sqrt{T}$

∴ v = k√T (where k is a constant)
Substituting v = 332ms⁻¹, and T = 25 + 273 = 298k, gives:
$$332 = k\sqrt{298}$$
∴ $k = \dfrac{332}{\sqrt{298}}$

When T = 37 + 273 = 310k, then v is obtained as follows:
$$v = k\sqrt{T}$$
$$= \dfrac{332}{\sqrt{298}} \times \sqrt{310} \quad (k = \dfrac{332}{\sqrt{298}} \text{ from above})$$
$$= 332\sqrt{\dfrac{310}{298}}$$
$$= 332 \times \sqrt{1.0403}$$
$$= 332 \times 1.02 = 338.6$$

The velocity at 37°C is 338.6ms⁻¹

8. The velocity of sound in air at a certain temperature is 344ms⁻¹. What is the velocity when the absolute temperature is increased by 16%?

Solution

If the initial temperature is T_1, then increasing T_1 by 16% gives,
$$T_1 + \dfrac{16}{100}T_1 = T_1 + 0.16T_1 = 1.16T_1$$
$$v \propto \sqrt{T}$$
∴ v = k√T (where k is a constant)
Substituting v = 344ms⁻¹, and T = T_1, gives,
$$344 = k\sqrt{T_1} \quad \ldots\ldots\ldots\ldots\text{Equation 1}$$
Substituting v = v_2, and T = T_2 = 1.16T_1 gives:
$$v_2 = k\sqrt{1.16T_1} \quad \ldots\ldots\ldots\ldots\text{Equation 2}$$
Equation 2 divided by equation 1 gives:
$$\dfrac{v_2}{344} = \dfrac{k\sqrt{1.16T_1}}{k\sqrt{T_1}}$$
$$\dfrac{v_2}{344} = \sqrt{\dfrac{1.16T_1}{T_1}} \quad (k \text{ cancels out})$$
$$\dfrac{v_2}{344} = \sqrt{1.16} \quad (T_1 \text{ cancels out})$$
$$\dfrac{v_2}{344} = 1.077$$
∴ v_2 = 344 x 1.077 = 370.5

The velocity when the temperature is increased by 16% is 370.5ms⁻¹

Exercise 5

1. A man stands in front of a mountain and fires a gun. He hears the echo from the mountain after 4sec. If the speed of sound in air is 330m/s, how far is the man from the mountain?

2. A horn is sounded from a tall building, 1200m way. How long will it take to hear the echo? (Speed of sound in air is 340m)

3. A whistle is blown at a distance of 332m from a vertical wall. If the echo of the sound produced by the whistle is heard after 2sec, calculate the speed of sound in air.

4. Two people P and Q are 220m apart along a horizontal ground. A cliff is 180m behind Q. P blows a whistle. What is the time interval between the two sounds:

(a) heard by Q

(b) heard by P?

(Speed of sound in air is 340m)

5. A horn is sounded at regular intervals in front of a vertical wall 200m away. If the reflected sound from the wall is heard simultaneously with the next hoot, how many hoots are made every minute. (Velocity of sound in air = $330 ms^{-1}$)

6. A boy 162m from a cliff blows a whistle once every 2 seconds. He hears the echo of each whistle midway between the whistle and the next whistle. Calculate the speed of sound in air.

7. The velocity of sound in air at a temperature of 20°C is $328 ms^{-1}$. What is the velocity when the temperature is 39°C?

8. The velocity of sound in air at a certain absolute temperature is $338 ms^{-1}$. What is the velocity when the absolute temperature is increased by 25%?

CHAPTER 6
BEAT

Beat is a phenomenon which occurs when there is a rise and fall in the loudness of sound when two notes of nearly equal frequency are sounded together.

If f_1 and f_2 are the frequencies of the two notes with f_1 greater than f_2 then the beat frequency is given by:

$$f = f_1 - f_2$$

The beat frequency also means the number of beats made per second.
The period of the beat is given by:

$$T = \frac{1}{f}$$

Examples

1. Two notes of frequencies 30Hz and 25Hz are sounded together. Calculate:
(a) the beat frequency
(b) the period of the beat

<u>Solution</u>

(a) $f = f_1 - f_2$ (Where f_1 is the larger frequency)
    = 30 – 25 = 5Hz
∴ $f = 5Hz$

(b) Recall that, $T = \frac{1}{f}$

∴ $T = \frac{1}{5} = 0.2\text{sec}$

The period of the beat is 0.2sec

2. Two notes sounded together produced a beat of period 0.25sec. If the lower note has a frequency of 452Hz, calculate:
(a) the beat frequency
(b) the frequency of the higher note

<u>Solutions</u>

(a) The period is given by:
$$T = \frac{1}{f}$$
∴ $f = \frac{1}{T}$
$= \frac{1}{0.25} = 4$

The beat frequency is 4Hz

(b) $f = f_1 - f_2$ Where f is the beat frequency and f_1 is the larger frequency of the two notes.

$\quad 4 = f_1 - 452$

$\therefore \quad 4 + 452 = f_1$

$\quad\quad 456 = f_1$

$\quad\quad f_1 = 456$

The frequency of the higher note is 456Hz

3. A note of 62Hz is sounded together with a note of 66Hz. How many beats per second will be heard?

Solution

The number of beats heard, $f = f_1 - f_2$

$\quad\quad\quad\quad = 66 - 62 = 4Hz$

The number of beats that will be heard is 4 beats per second.

Exercise 6

1. Two notes of frequencies 440Hz and 444Hz are sounded together. Calculate:
(a) the beat frequency
(b) the period of the beat

2. Two notes sounded together produced a beat of period 0.8sec. If the lower note has a frequency of 52Hz, calculate:
(a) the beat frequency
(b) the frequency of the higher note

3. A note of 100Hz is sounded together with a note of 106Hz. How many beats per second will be heard?

4. A beat per second of 3Hz was heard after sounding two notes. If the note of one of the sounds is 20Hz, calculate the two possible values of the other note.

5. Two notes sounded together produced a beat of period 1.4sec. If the higher note has a frequency of 325Hz, calculate:
(a) the beat per second heard
(b) the frequency of the lower note

CHAPTER 7
VIBRATION OF AIR COLUMN IN PIPES

Vibration in a closed pipe

At first resonance the length of air column in a closed pipe is given by:
$$l + c = \lambda/4$$
where l is the distance of the water level below the pipe, and c is the end correction which is a little distance of the wave above the top of the pipe. The value of c is usually negligible thereby giving the length of the air column as:
$$l = \lambda/4$$
Or, $\quad \lambda = 4l$

Therefore the resonant frequency of vibration is:
$$f_0 = v/\lambda \quad \text{(from } v = f\lambda\text{)}$$
$$= \frac{v}{4l}, \text{ where v is the speed of sound in air, } f_0 \text{ is the fundamental frequency of}$$
the closed pipe.

The length of air column where the next resonance is observed in a closed pipe is given by:
$$l = 3\lambda/4$$
Or, $\quad \lambda = \dfrac{4l}{3}$

Therefore the frequency of vibration is given by:
$$f_1 = v/\lambda$$
$$= v \div \frac{4l}{3} \quad \text{(Since } \lambda = \frac{4l}{3}\text{)}$$
$$f_1 = \frac{3v}{4l}$$
$$f_1 = 3f_0 \quad \text{(Since } f_0 = \frac{v}{4l}\text{)}$$

This frequency, $f_1 = 3f_0$, is called the third harmonic or first overtone of a closed pipe. f_0 is the first harmonic. Other frequencies that can be obtained in a closed pipe include: $f_2 = 5f_0$, which is the fifth harmonic or second overtone, $f_3 = 7f_0$ and so on.

Vibration in an open pipe

At first resonance the length of air column in a open pipe is given by:
$$l = \lambda/2$$
Or, $\quad \lambda = 2l$

Therefore the resonant frequency of vibration is:
$$f_0 = \frac{v}{2l},$$
The length of air column where the next resonance is observed (i.e. at the second harmonic or first overtone) in an open pipe is given by:

$$l = \lambda$$

Therefore the frequency of vibration is given by:

$$f_1 = \frac{v}{l}$$

$$= 2(\frac{v}{2l}), \quad \text{(Note that } 2(\frac{v}{2l}) = \frac{v}{l})$$

$$f_1 = 2f_0 \quad \text{(Since } \frac{v}{2l} = f_0)$$

Similarly, the third harmonic, $f_2 = 3f_0$.

This shows that for an open pipe, $f_1 = 2f_0$, which is the second harmonic or first overtone, $f_2 = 3f_0$, which is the third harmonic or second overtone and so on. Therefore all harmonics are possible in an open pipe.

Note that: At first overtone, $l = \lambda$, at second overtone, $l = 3\lambda/2$, at third overtone, $l = 2\lambda$, and so on.

Examples

1. Calculate the frequency of fundamental of a closed pipe of length 20cm, if the speed of sound in air is 340m/s

Solution

$l = \lambda/4$ (The length of the first harmonic of a closed pipe, where λ is the wavelength)

$\therefore \lambda = 4l$

But, $l = 20cm = (\frac{20}{100})$ m = 0.2m

$\lambda = 4 \times 0.2 = 0.8m$

Recall that, $v = f\lambda$

$\therefore f = v/\lambda$

$= \frac{340}{0.8} = 425$

$f = 425Hz$

2. If the length of the air column in a closed pipe is 45cm when it experiences its first overtone, the wavelength of the note is what? Calculate the frequency of the second overtone if the speed of sound in air is 340m/s

Solution

$l = 3\lambda/4$ (The length of the first overtone of a closed pipe)

$\therefore \lambda = \frac{4l}{3}$

But, $l = 45cm = (\frac{45}{100})$ m = 0.45m

$\lambda = \frac{4 \times 0.45}{3}$

$$= \frac{1.8}{3}$$
$$= 0.6m$$

The first overtone wavelength of the note is 0.6m

Recall that, $f = v/\lambda$ (Where $f = f_1$ = the frequency of the first overtone)

$$\therefore f_1 = v/\lambda$$
$$= \frac{340}{0.6} = 566.7$$
$$f_1 = 566.7Hz$$

But, $f_1 = 3f_0$

$$\therefore f_0 = \frac{f_1}{3}$$
$$= \frac{566.7}{3}$$
$$f_0 = 188.9Hz$$

But, $f_2 = 5f_0$ (Where f_2 is the frequency of the second overtone)

$$\therefore f_2 = 5 \times 188.9 = 944.5$$

The frequency of the second overtone $f_2 = 944.5Hz$

3. If the length of the air column in a closed pipe is 60cm when it experiences its third overtone, what is the wavelength of the note? Calculate the frequency of the first overtone if the speed of sound in air is 340m/s

Solution

$l = 7\lambda/4$ (The length of the third overtone of a closed pipe)

$$\therefore \lambda = \frac{4l}{7}$$

But, $l = 60cm = (\frac{60}{100}) m = 0.60m$

$$\lambda = \frac{4 \times 0.6}{7}$$
$$= \frac{2.4}{7}$$
$$= 0.343m$$

The third overtone wavelength of the note is 0.343m

Recall that, $f = v/\lambda$ (Where $f = f_3$ = the frequency of the third overtone)

$$\therefore f_3 = v/\lambda$$
$$= \frac{340}{0.343} = 991.3$$
$$f_3 = 991.3Hz$$

But, $f_3 = 7f_0$

$$\therefore f_0 = \frac{f_3}{7}$$

$$= \frac{991.3}{7}$$

$f_0 = 141.6\text{Hz}$

But, $f_1 = 3f_0$ (Where f_1 is the frequency of the first overtone)

$\therefore \quad f_1 = 3 \times 141.6 = 424.8$

The frequency of the first overtone $f_1 = 424.8\text{Hz}$

4. The length of the air column in an open pipe is 108cm when it attains its first overtone. Calculate the frequency of the third overtone if the speed of sound in air is 330m/s

Solution

$l = \lambda$ (The length of the first overtone of an open pipe)

$\therefore \lambda = l$

But, $l = 108\text{cm} = \left(\frac{108}{100}\right)\text{m} = 1.08\text{m}$

$\lambda = 1.08\text{m}$

Recall that, $f = v/\lambda$ (Where $f = f_1$ = the frequency of the first overtone)

$\therefore f_1 = v/\lambda$

$$= \frac{330}{1.08} = 305.6$$

$f_1 = 305.6\text{Hz}$

But, $f_1 = 2f_0$ (First overtone of an open tube)

$\therefore \quad f_0 = \frac{f_1}{2}$

$$= \frac{305.6}{2}$$

$f_0 = 152.8\text{Hz}$

But, $f_3 = 4f_0$ (Where f_3 is the frequency of the third overtone)

$\therefore \quad f_3 = 4 \times 152.8 = 611.2$

The frequency of the third overtone $f_3 = 611.2\text{Hz}$

5. What is the shortest length of an opened tube which resonates with a frequency of 320Hz? (speed of sound in air = 330m/s)

Solution

$l = \lambda/2$ (Shortest length is the length of the first harmonics)

$\therefore \lambda = 2l$

But, $v = f\lambda$

$v = f \times 2l$ (Since $\lambda = 2l$)

$\therefore 2fl = v$

$$l = \frac{v}{2f}$$

$$l = \frac{330}{2 \times 320}$$
$$= \frac{330}{640} = 0.516m$$

6. The length of the vibrating air column of a simple resonance tube can be altered by adjusting a water level. Resonance is found for a tuning fork of frequency 440Hz when the length of the air column is 18.8cm and again when it is 57.3cm. Calculate the speed of sound of the air in the tube.

Solution

Assuming the two positions to be the first and second resonance positions, then:

$l_1 = \lambda/4$ (Position of first resonance)
$l_2 = 3\lambda/4$ (Position of second resonance)
$l_2 - l_1 = 3\lambda/4 - \lambda/4$
$l_2 - l_1 = 2\lambda/4$
$l_2 - l_1 = \lambda/2$ (When expressed in its lowest term)
$57.3 - 18.8 = \lambda/2$
$38.5 = \lambda/2$
∴ $\lambda = 2 \times 38.5 = 77cm$
$\lambda = (\frac{77}{100})m = 0.77m$

But, $v = f\lambda$
$v = 440 \times 0.77$ (f = 440, as given in the question)
$v = 338.8 ms^{-1}$

The speed of sound of the air in the tube is $338.8 ms^{-1}$

7. In a resonance tube experiment, resonance is found for a tuning fork of frequency 480Hz when the length of the air column is 24.7cm and again when the air column is 59.9cm long. Calculate the speed of sound in air.

Solution

As shown in example (6) above, the difference in length of air column between any two consecutive positions of resonance in a closed pipe is $\lambda/2$.

(e.g. $5\lambda/4 - 3\lambda/4 = \lambda/2$)

∴ $59.9 - 24.7 = \lambda/2$
$35.2 = \lambda/2$
∴ $\lambda = 2 \times 35.2 = 70.4cm$
$\lambda = (\frac{70.4}{100})m = 0.704m$

But, $v = f\lambda$

∴   v = 480 x 0.704   (f = 480, as given in the question)
    v = 337.9ms^{-1}

The speed of sound in air is 337.9ms^{-1}

Exercise 7

1. Calculate the frequency of fundamental of a closed pipe of length 42cm, if the speed of sound in air is 330m/s

2. If the length of the air column in an open pipe is 45cm when it experiences its first overtone, what is the wavelength of the note? Calculate the frequency of the third overtone if the speed of sound in air is 340m/s

3. If the length of the air column in a closed pipe is 25cm when it experiences its second overtone, what is the wavelength of the note? Calculate the frequency of the first overtone if the speed of sound in air is 340m/s

4. The length of the air column in an open pipe is 75cm when it attains its second overtone. Calculate the frequency of the fifth overtone if the speed of sound in air is 330m/s

5. What is the shortest length of an opened tube which resonates with a frequency of 82Hz? (Speed of sound in air = 330m/s)

6. The length of the vibrating air column of a simple resonance tube can be altered by adjusting a water level. Resonance is found for a tuning fork of frequency 380Hz when the length of the air column is 24.2cm and again when it is 72.1cm. Calculate the speed of sound of the air in the tube.

7. In a resonance tube experiment, resonance is found for a tuning fork of frequency 560Hz when the length of the air column is 16.5cm and again when the air column is 46.3cm long. Calculate the speed of sound in air.

8. The length of the air column in an open pipe is 0.96m when it attains its third overtone. Calculate the frequency of the first overtone if the speed of sound in air is 340m/s

CHAPTER 8
MODES OF VIBRATION OF A STRETCHED STRING

The simplest mode of vibration (fundamental) of a plucked stretched string is given by:
$$l = \lambda/2$$
$$\therefore \lambda = 2l$$
The fundamental frequency is thus given by:
$$f_0 = \frac{v}{2l} \quad \text{(from } v = f\lambda\text{)}$$
At first overtone, $l = \lambda$, at second overtone, $l = 3\lambda/2$, at third overtone, $l = 2\lambda$, and so on.
As with an open pipe, all possible harmonics, f_0, $f_1=2f_0$, $f_2=3f_0$, $f_3=4f_0$, etc, can be obtained on a stretched string.
The velocity of a wave propagated along a fixed wire or string is given by:
$$v = \sqrt{\frac{T}{m}}$$
The fundamental frequency can therefore be given by:
$$f_0 = \frac{1}{2l}\sqrt{\frac{T}{m}} \quad \text{(from } f = v/\lambda\text{), where T is the tension in the string,}$$
m is the mass per unit length of the string in kg/m, and l is the length of the string in metre. This shows that:

$f \alpha \sqrt{T}$ (when l and m are constant)

$f \alpha \frac{1}{l}$ (when T and m are constant)

$f \alpha \frac{1}{\sqrt{m}}$ (when T and l are constant)

Examples

1. The frequency of a standard string when the tension on the string is 20N was found to be 45Hz. Determine the frequency when the tension is increased to 80N.

<u>Solution</u>

$f \alpha \sqrt{T}$ (f is the frequency while T is the tension)

$\therefore f = k\sqrt{T}$ (Where k is a constant)

When f = 45 and T = 20, then:

$45 = k\sqrt{20}$

$k = \frac{45}{\sqrt{20}}$

When T = 80, then f is obtained as follows:

$f = k\sqrt{T}$

$$= \frac{45}{\sqrt{20}} \times \sqrt{80} \quad \text{(Since } k = \frac{45}{\sqrt{20}}\text{)}$$

$$= 45\sqrt{\frac{80}{20}}$$

$$= 45\sqrt{4}$$

$$= 45 \times 2 = 90$$

The frequency is 90Hz

2. The frequency of a guitar string when the tension on the string is 42N was found to be 60Hz. Determine the frequency of the string when the tension is decreased to 12N.

Solution

$f \alpha \sqrt{T}$

$\therefore f = k\sqrt{T}$ (Where k is a constant)

When f = 60 and T = 42, then:

$60 = k\sqrt{42}$

$k = \frac{60}{\sqrt{42}}$

When T = 12, then f is obtained as follows:

$f = k\sqrt{T}$

$= \frac{60}{\sqrt{42}} \times \sqrt{12} \quad \text{(Since } k = \frac{60}{\sqrt{42}}\text{)}$

$= 60\sqrt{\frac{12}{42}}$

$= 60\sqrt{0.2857}$

$= 60 \times 0.5345$

$= 32.1Hz$

3. A string has a frequency of 54Hz when a tension of 28N is applied on it. Determine the new tension on the string when the frequency is 20Hz.

Solution

Another method of solving a question like this is as explained below.

$f = k\sqrt{T}$ (Where k is a constant)

When f = 54 and T = 28, then:

$54 = k\sqrt{28}$ .... Equation (1)

When f = 20 and T = T_2, then:

$20 = k\sqrt{T_2}$ .... Equation (2) (T_2 is the unknown tension to be calculated)

Divide equation (1) by equation (2). This gives:

$$\frac{54}{20} = \frac{k\sqrt{28}}{k\sqrt{T2}}$$

(k cancels out to give)

$$\therefore \frac{54}{20} = \sqrt{\frac{28}{T2}}$$

$$2.7 = \sqrt{\frac{28}{T2}}$$

Square both sides. This gives:

$$2.7^2 = \frac{28}{T2}$$ (When $\sqrt{\frac{28}{T2}}$ is squared, the square root sign clears out)

$$\therefore 7.29 = \frac{28}{T2}$$

$$7.29 T_2 = 28$$

$$T_2 = \frac{28}{7.29} = 3.84$$

The new tension on the string is 3.84N

4. The frequency of a string emitting a note is 50Hz. Determine its frequency when the tension is halved.

Solution

Let the original and new tensions be T_1 and T_2 respectively. (But $T_2 = \frac{T_1}{2}$ since the new tension is half of the original tension)

$f = k\sqrt{T}$ (Where k is a constant)

When f = 50 and T = T_1, then:

$50 = k\sqrt{T_1}$ .... Equation (1)

When f = f_2 and T = $T_2 = \frac{T_1}{2}$, then:

$f_2 = k\sqrt{\frac{T_1}{2}}$ .... Equation (2)   (f_2 is the unknown frequency to be calculated)

Divide equation (1) by equation (2). This gives:

$$\frac{50}{f_2} = \frac{k\sqrt{T_1}}{k\sqrt{\frac{T_1}{2}}}$$

$$\therefore \frac{50}{f_2} = \sqrt{\frac{T_1}{\frac{T_1}{2}}}$$ (k cancels out)

$$\frac{50}{f_2} = \sqrt{\frac{2T_1}{T_1}}$$

$$\frac{50}{f_2} = \sqrt{2} \quad (T_1 \text{ cancels out})$$

$$\frac{50}{f_2} = 1.414$$

∴ $1.414f_2 = 50$

$$f_2 = \frac{50}{1.414} = 35.4$$

The frequency is 35.4N

5. A string emits a note of frequency 100Hz when a certain tension was applied on it. Determine the frequency of the string when the tension is quadrupled.

Solution

Let the original and new tensions be T_1 and T_2 respectively. (But $T_2 = 4T_1$ since the new tension is four times the original tension i.e. quadrupled.

$f = k\sqrt{T}$

When $f = 100$ and $T = T_1$, then:

$100 = k\sqrt{T_1}$

∴ $k = \frac{100}{\sqrt{T_1}}$

When $f = f_2$ and $T = T_2 = 4T_1$, then:

$f_2 = k\sqrt{T_2}$

$f_2 = \frac{100}{\sqrt{T_1}} \times \sqrt{4T_1}$ (Since $k = \frac{100}{\sqrt{T_1}}$ and $T_2 = 4T_1$)

$f_2 = 100\sqrt{\frac{4T_1}{T_1}}$

T_1 cancels out to give:

$f_2 = 100\sqrt{4}$

$f_2 = 100 \times 2 = 200$

The frequency of the string is 200Hz

6. A string of length 42cm and mass 12g emits a note when a tension of 75N was applied on it. Determine the fundamental frequency of the string.

Solution

Given: T = 75N, l = 42cm = $(\frac{42}{100})$m = 0.42m

m = mass per unit length. But mass of wire = $(\frac{12}{1000})$kg = 0.012kg

∴ mass per unit length, m = $\frac{0.012}{0.42}$ = 0.02857kg/m

So, the fundamental frequency is given by:

$$f_0 = \frac{1}{2l}\sqrt{\frac{T}{m}}$$

$$= \frac{1}{2 \times 0.42}\sqrt{\frac{75}{0.02857}}$$

$$= \frac{1}{0.84}\sqrt{2625}$$

$$= \frac{1}{0.84} \times 51.23 = \frac{51.23}{0.84} = 60.988$$

The fundamental frequency, f_0 is 61.0Hz

7. A string of length 80cm and mass 4g emits a note when a tension of 20N was applied on it. Determine:
(a) the frequency of the third harmonic
(b) the frequency of the fifth overtone
(c) the velocity of the wave in the string

Solution

(a) An easy method of solving this question is to first determine the fundamental frequency, f_0.

Given: T = 20N, l = 80cm = $(\frac{80}{100})$m = 0.8m

m = mass per unit length. But mass of wire = $(\frac{4}{1000})$kg = 0.004kg

∴ mass per unit length, m = $\frac{0.004}{0.8}$ = 0.005kg/m

So, the fundamental frequency is given by:

$$f_0 = \frac{1}{2l}\sqrt{\frac{T}{m}}$$

$$= \frac{1}{2 \times 0.8}\sqrt{\frac{20}{0.005}}$$

$$= \frac{1}{1.6}\sqrt{4000}$$

$$= \frac{1}{1.6} \times 63.25$$

$$= \frac{63.25}{1.6} = 39.5$$

The fundamental frequency, f_0 is 39.5Hz

∴ The frequency of the third harmonic = $3f_0$ = 3 × 39.5 = 118.5 (Note that harmonics are multiples of the fundamental frequency)
Hence, the frequency of the third harmonic is 118.5Hz

(b) The frequency of the fifth overtone, $f_5 = 6f_0$

∴ $f_5 = 6 \times 39.5 = 237$

The frequency of the fifth overtone is 237Hz.

(c) The velocity of the wave is given by:

$$v = \sqrt{\frac{T}{m}}$$

$$= \sqrt{\frac{20}{0.005}}$$

$$= \sqrt{4000}$$

$$= 63.25$$

∴ The velocity of the wave is 63.25ms^{-1}

8. A string when plucked emits a note of frequency 36Hz. Determine the frequency of the string when the length is tripled, and the tension on the string is kept constant.

Solution

Let the original and new lengths be l_1 and l_2 respectively. (But $l_2 = 3l_1$ since the new length is three times the original length)

∴ $f \alpha \dfrac{1}{l}$ (when T and m are constant)

$f = \dfrac{k}{l}$ (where k is a constant)

When $f = f_1 = 36$Hz, and $l = l_1$ then f_1 is given by:

$f_1 = \dfrac{k}{l_1}$

$36 = \dfrac{k}{l_1}$

∴ $k = 36l_1$

When $f = f_2$, and $l = l_2 = 3l_1$ then f_2 is given by:

$f_2 = \dfrac{k}{l_2}$

$= \dfrac{36l_1}{3l_1}$ (Since $k = 36l_1$)

$= 12$ (l_1 cancels out)

The frequency is 12Hz when the length is tripled.

9. When a musical string is plucked, a note of frequency 120Hz is emitted. Determine the frequency of the string when its mass per unit length is halved, while the tension on the string and length of the string are kept constant.

Solution

Let the original and new mass per unit engths be m_1 and m_2 respectively. (But $m_2 = \frac{m_1}{2}$ since the new mass per unit length is half times the original value)

∴ $f \alpha \frac{1}{\sqrt{m}}$ (when T and l are constant)

$f = \frac{k}{\sqrt{m}}$ (where k is a constant)

When $f = f_1 = 120$Hz, and $m = m_1$ then f_1 is given by:

$f_1 = \frac{k}{\sqrt{m_1}}$

$120 = \frac{k}{\sqrt{m_1}}$

∴ $k = 120\sqrt{m_1}$

When $f = f_2$, and $m = m_2 = \frac{m_1}{2}$ then f_2 is given by:

$f_2 = \frac{k}{\sqrt{m_1}}$

$= \frac{120\sqrt{m_1}}{\sqrt{\frac{m_1}{2}}}$ (Since $k = 120\sqrt{m_1}$)

$= 120\sqrt{m_1} \times \sqrt{\frac{2}{m_1}}$

$= 120 \times \sqrt{\frac{2m_1}{m_1}}$

$= 120 \times \sqrt{2}$ (m_1 cancels out)

$= 120 \times 1.414$

$= 169.7$

The frequency is 169.7Hz.

10. A plucked string of length 70cm emits a note of certain frequency. Determine the new length of the string when the frequency is increased to five times its original value, and the tension on the string is kept constant.

Solution

Let the original and new frequencies be f_1 and f_2 respectively. (But $f_2 = 5l_1$ since the new frequency is five times the original frequency)

∴ $f \alpha \frac{1}{l}$ (when T and m are constant)

$f = \frac{k}{l}$ (where k is a constant)

When $f = f_1$, and $l = l_1 = 70$cm then f_1 is given by:

$$f_1 = \frac{k}{l_1}$$

$$f_1 = \frac{k}{70}$$

$\therefore k = 70f_1$

When $f = f_2 = 5f_1$, and $l = l_2$, then f_2 is given by:

$$f_2 = \frac{k}{l_2}$$

$$5f_1 = \frac{70f_1}{l_2} \quad \text{(Since } k = 70f_1 \text{ and } f_2 = 5f_1\text{)}$$

$$5f_1 l_2 = 70f_1$$

$$\therefore l_2 = \frac{70f_1}{5f_1}$$

$= 14$ (f_1 cancels out)

The new length is 14Hz when the frequency is increased five times.

Exercise 8

1. The frequency of a standard string when the tension on the string is 32N was found to be 50Hz. Determine the frequency when the tension is increased to 60N.

2. The frequency of a violin string when the tension on the string is 100N was found to be 85Hz. Determine the frequency of the string when the tension is increased to 140N.

3. A string has a frequency of 300Hz when a tension of 120N is applied on it. Determine the new tension on the string when the frequency becomes 80Hz.

4. The frequency of a string emitting a note is 50Hz. Determine its frequency when the tension becomes five times its original value.

5. A string emits a note of frequency 420Hz when a certain tension was applied on it. Determine the frequency of the string when the tension is halved.

6. A string of length 0.65m and mass 10g emits a note when a tension of 100N was applied on it. Determine the fundamental frequency of the string.

7. A string of length 210cm and mass 20g emits a note when a tension of 32N was applied on it. Determine:

(a) the frequency of the second harmonic

(b) the frequency of the fourth overtone

(c) the velocity of the wave in the string

8. A string when plucked emits a note of frequency 220Hz. Determine the frequency of the string when the length is quadrupled, and the tension on the string remains the same.

9. When a musical string is plucked, a note of frequency 25Hz is emitted. Determine the frequency of the string when its mass per unit length is tripled, while the tension on the string and length of the string are kept constant.

10. A plucked string of length 100cm emits a note of certain frequency. Determine the new length of the string when the frequency is doubled and the tension on the string is kept constant.

CHAPTER 9
CHARACTERISTICS OF SOUND – THE PITCH

The pitch of a sound depends on the frequency of the sound wave. Two methods which can be used to show that the pitch of a sound depends on its frequency are:

1. The use of disc siren. This method shows that the frequency of the wave produced is given by:

$$f = \frac{\text{no. of revolutions} \times \text{no. of holes on the disc}}{\text{time of revolution in seconds}}$$

2. The use of toothed wheel. This method shows that the frequency of the wave produced is given by:

$$f = \frac{\text{no. of revolutions} \times \text{no. of teeth on the wheel}}{\text{time of revolution in seconds}}$$

Examples

1. An experiment was carried out to determine the frequency of a note. A disc siren was observed to make a revolution of 360 for a period of 3 minutes. If the disc is made up of 40 evenly spaced holes, what is the frequency of the note emitted?

<u>Solution</u>

$$f = \frac{\text{no. of revolutions} \times \text{no. of holes on the disc}}{\text{time of revolution in seconds}}$$

$$= \frac{360 \times 40}{3 \times 60} \quad \text{(The 3 minutes is multiplied by 60 to convert it to seconds)}$$

$$= 20 \times 4 \quad \text{(After equal divisions by 10, 6 and then 3)}$$

$$\therefore f = 80 \text{ vibrations per second or 80Hz}$$

2. A disc siren produced a note of frequency 240Hz. If the disc is made up of 60 evenly spaced holes, what is the speed of the disc?

<u>Solution</u>

Let the speed be the number of revolutions made per second, i.e. in 1 second. So, the time is 1 second.

$$f = \frac{\text{no. of revolutions} \times \text{no. of holes on the disc}}{\text{time of revolution in seconds}}$$

$$240 = \frac{\text{no. of revolutions} \times 60}{1}$$

$$240 = \frac{\text{no. of revolutions}}{1} \times 60$$

$$\therefore \text{No. of revolutions/sec} = \frac{240}{60}$$

$$= 4$$

(Note that $\frac{\text{no. of revolutions}}{1}$ is the no of revolutions/sec)

∴ Speed of the disc is 4revs./sec

3. In an experiment to produce a note of frequency 300Hz, a toothed wheel was rotated to make 250 revolutions in 2 minutes. Calculate the number of teeth on the wheel.

Solution

$f = \frac{\text{no. of revolutions} \times \text{no. of teeth on the wheel}}{\text{time of revolution in seconds}}$

$300 = \frac{250 \times \text{no. of teeth}}{2 \times 60}$ (The 2 minutes is multiplied by 60 to convert it to seconds)

$300 \times 2 \times 60 = 250 \times$ no. of teeth

∴ No. of teeth $= \frac{300 \times 2 \times 60}{250}$

$= 12 \times 2 \times 6$ (After equal divisions by 10 and 25)

$= 144$

The wheel has 144 teeth.

4. A note of frequency 200Hz was produced by a toothed wheel. If the wheel has 60 teeth, calculate the time taken by the wheel to complete 240 revolutions.

Solution

$f = \frac{\text{no. of revolutions} \times \text{no. of teeth on the wheel}}{\text{time of revolution in seconds}}$

$200 = \frac{240 \times 60}{\text{time in seconds}}$

$200 \times \text{time} = 240 \times 60$

∴ Time $= \frac{240 \times 60}{200}$

$= 12 \times 6$ (After cancelling out the zeros and dividing by 2)

$= 72$ seconds

Time in minutes $= \frac{72}{60} = 1.2$ minutes

It takes 1.2 minutes for the wheel to make the revolutions.

Exercise 9

1. An experiment was carried out to determine the frequency of a note. A disc siren was observed to make a revolution of 540 for a period of 4 minutes. If the disc is made up of 60 evenly spaced holes, what is the frequency of the note emitted?

2. A disc siren produced a note of frequency 44Hz. If the disc is made up of 50 evenly spaced holes, what is the speed of the disc?

3. In an experiment to produce a note of frequency 80Hz, a toothed wheel was rotated to make 320 revolutions in 2.5 minutes. Calculate the number of teeth on the wheel.

4. A note of frequency 540Hz was produced by a toothed wheel. If the wheel has 80 teeth, calculate the time taken by the wheel to complete 150 revolutions.

5. A note of frequency 100Hz was produced by a disc siren. If the disc has 40 evenly spaced holes, calculate the time taken by the disc to complete 380 revolutions.

CHAPTER 10
DOPPLER EFFECTS IN SOUND

Doppler effect is the change in frequency (pitch) of a source of sound when there is a relative motion between the source and an observer.

Expressions for apparent frequencies

Case 1 When the source of sound is moving towards a stationary observer, the apparent frequency is given by:

$$f' = \left(\frac{V}{V-V_s}\right)f$$

Where f' is the apparent frequency, V is the velocity of sound in air, V_s is the velocity of the source of sound, and f is the frequency emitted from the source of sound.

Case 2 When the source of sound is moving away from a stationary observer, the apparent frequency is given by:

$$f' = \left(\frac{V}{V+V_s}\right)f$$

Case 3 When the source of sound is stationary and an observer is moving towards it, then the apparent frequency is given by:

$$f' = \left(\frac{V+V_o}{V}\right)f \quad \text{where } V_o \text{ is the velocity of the observer.}$$

Case 4 When the source of sound is stationary and an observer is moving away from it, then the apparent frequency is given by:

$$f' = \left(\frac{V-V_o}{V}\right)f$$

Case 5 When both the source of sound and an observer are moving towards each other, then the apparent frequency is given by:

$$f' = \left(\frac{V+V_o}{V-V_s}\right)f$$

Case 6 When both the source of sound and an observer are moving away from each other, then the apparent velocity is given by:

$$f' = \left(\frac{V-V_o}{V+V_s}\right)f$$

Case 7 When the source of sound and an observer are moving in the same direction, then we can have two cases as stated below.

(a) When the source is behind the observer, then the apparent frequency is given by:
$$f' = \left(\frac{V-V_o}{V-V_s}\right)f$$

(b) When the source is in front of the observer, then the apparent frequency is given by:
$$f' = \left(\frac{V+V_o}{V+V_s}\right)f$$

Note that the quantities, $V-V_o$, $V+V_o$, $V-V_s$ and $V+V_s$ are the relative velocities between the velocity of sound in air and the velocity of the source of sound or the observer. When both velocities are moving in the same direction we subtract their velocities. When they are moving in opposite direction we add their velocities. All sounds are moving away from the source to the observer.

Examples

1. A train approaching a station with a speed of 25ms^{-1} sounds its horn and emits a note of frequency 450Hz. What frequency will be received at the station? (Speed of sound in air is 330ms^{-1}).

Solution

This is case 1 as stated above. The apparent frequency is given by:
$$f' = \left(\frac{V}{V-V_s}\right)f$$

where V=330ms^{-1}, V_s = 25ms^{-1} and f = 450Hz. Substituting these values into the expression above gives:

$$f' = \left(\frac{330}{330-25}\right) \times 450$$
$$= \left(\frac{330}{305}\right) \times 450$$
$$= 486.9$$

The station will receive a frequency of 486.9Hz.

2. A sounding tuning fork of frequency 420Hz is moved away from an observer with a speed of 4ms^{-1}. What is the apparent frequency of the sound coming to the observer? (Speed of sound in air is 330ms^{-1}).

Solution

This is case 2 above. The apparent frequency is given by:
$$f' = \left(\frac{V}{V+V_s}\right)f$$

where V=330ms^{-1}, V_s = 4ms^{-1} and f = 420Hz. Substituting these values into the expression above gives:

$$f' = \left(\frac{330}{330+4}\right) \times 420$$

$$= (\frac{330}{334}) \times 420$$
$$= 415$$

The frequency coming to the observer is 415Hz.

3. A stationary siren emits a note of frequency 382Hz. What frequency will be received by a train which is approaching it at a speed of 20ms^{-1}? (Speed of sound in air is 330ms^{-1}).

Solution

This is case 3 above. The apparent frequency is given by:

$$f' = (\frac{V+V_o}{V})f$$

where V=330ms^{-1}, V$_o$ = 20ms^{-1} and f = 382Hz. Substituting these values into the expression above gives:

$$f' = (\frac{330+20}{330}) \times 382$$
$$= (\frac{350}{330}) \times 382$$
$$= 405.2$$

The train will receive a frequency of 405.2Hz.

4. A stationary siren at a train station emits a note of frequency 365Hz. What frequency will be received by a train which is leaving the station at a speed of 11ms^{-1}? (Speed of sound in air is 333ms^{-1}).

Solution

This is case 4 as stated above. The apparent frequency is given by:

$$f' = (\frac{V-V_o}{V})f$$

where V=333ms^{-1}, V$_o$ = 11ms^{-1} and f = 365Hz. Substituting these values into the expression above gives:

$$f' = (\frac{333-11}{333}) \times 365$$
$$= (\frac{322}{333}) \times 365$$
$$= 352.9$$

The train will receive a frequency of 352.9Hz.

5. A student sounding a tuning fork of frequency 500Hz moves towards a man with a speed of 2ms^{-1}. If the man is running towards the student with a speed of 3ms^{-1}, what is the apparent frequency of the sound coming to the man? (Speed of sound in air is 333ms^{-1}).

Solution

This is case 5 above. The apparent frequency is given by:

$$f' = (\frac{V+V_o}{V-V_s})f$$

where V=333ms⁻¹, V_s = 2ms⁻¹, V_o = 3ms⁻¹ and f = 500Hz. Substituting these values into the expression above gives:

$$f' = (\frac{333+3}{333-2}) \times 500$$
$$= (\frac{336}{331}) \times 500$$
$$= 507.6$$

The frequency coming to the man is 507.6Hz.

6. A football referee and a footballer run away from each other on a football field. The referee runs with a velocity of 2ms⁻¹ while the footballer runs with a velocity of 4ms⁻¹. If the referee blows his whistle and emits a note of frequency 340Hz, what is the apparent frequency of the sound that the footballer will hear? (Speed of sound in air is 331ms⁻¹).

Solution

This is case 6 above. The apparent frequency is given by:

$$f' = (\frac{V-V_o}{V+V_s})f$$

where V=331ms⁻¹, V_s = 2ms⁻¹, V_o = 4ms⁻¹ and f = 340Hz. Substituting these values into the expression above gives:

$$f' = (\frac{331-4}{331+2}) \times 340$$
$$= (\frac{327}{333}) \times 340$$
$$= 333.9$$

The frequency that the footballer will hear is 333.9Hz.

7. A football referee and a footballer are running in the same direction with the referee behind the footballer. The referee runs with a velocity of 3ms⁻¹ while the footballer runs with a velocity of 2ms⁻¹. If the referee blows his whistle and emits a note of frequency 384Hz, what is the frequency of the sound coming to the footballer? (Speed of sound in air is 331ms⁻¹).

Solution

This is case 7a above. The apparent frequency is given by:

$$f' = (\frac{V-V_o}{V-V_s})f$$

where V=331ms⁻¹, V_s = 3ms⁻¹, V_o = 2ms⁻¹ and f = 384Hz. Substituting these values into the expression above gives:

$$f' = (\frac{331-2}{331-3}) \times 384$$

$$= (\frac{329}{328}) \times 384$$
$$= 385.2$$

The frequency that the footballer will hear is 385.2Hz.

8. A car travelling at a speed of 30ms^{-1} passes a bicycle which is moving at a speed of 7ms^{-1} in the opposite direction. If the car immediately sounds its horn and emits a note of frequency 480Hz, calculate the apparent frequency that will come to the bicycle? (Speed of sound in air is 331ms^{-1}).

Solution

This is case 7b above. The apparent frequency is given by:
$$f' = (\frac{V+Vo}{V+Vs})f$$
where V=331ms^{-1}, V$_s$ = 30ms^{-1}, V$_o$ = 7ms^{-1} and f = 480Hz. Substituting these values into the expression above gives:
$$f' = (\frac{331+7}{331+30}) \times 480$$
$$= (\frac{338}{361}) \times 480$$
$$= 449.4$$

The frequency that will come to the bicycle is 449.4Hz.

9. A stationary siren emits a note of frequency 430Hz. If the apparent frequency received by a train which is approaching it at a speed of 12.5ms^{-1} is 446Hz, determine the speed of sound in air.

Solution

This is case 3 above. The apparent frequency is given by:
$$f' = (\frac{V+Vo}{V})f$$
where f'= 446Hz, V$_o$ = 12.5ms^{-1} and f = 430Hz. Substituting these values into the expression above gives:
$$446 = (\frac{V+12.5}{V}) \times 430$$

Dividing both sides by 430 gives:
$$\frac{446}{430} = (\frac{V+12.5}{V})$$
$$1.037 = \frac{V+12.5}{V}$$
$$1.037V = V + 12.5$$
$$1.037V - V = 12.5$$
$$0.037V = 12.5$$

∴ $V = \dfrac{12.5}{0.037} = 337.8$

The speed of sound in air is 337.8ms^{-1}.

10. A student moving towards his classmate blows a whistle and emits a note of frequency 250Hz. If his classmate running towards him at a speed of 5ms^{-1}, hears the sound at an apparent frequency of 256Hz, determine the velocity with which the student was moving towards his classmate. (Speed of sound in air is 330ms^{-1}).

Solution

This is case 5 above. The apparent frequency is given by:

$$f' = \left(\dfrac{V+Vo}{V-Vs}\right)f$$

where f' = 256Hz, V=330ms^{-1}, V_o = 5ms^{-1} and f = 250Hz. Substituting these values into the expression above gives:

$$256 = \left(\dfrac{330+5}{330-Vs}\right) \times 250$$

Dividing both sides by 250 gives:

$$\dfrac{256}{250} = \left(\dfrac{335}{330-Vs}\right)$$

$$1.024 = \dfrac{335}{330-Vs}$$

$1.024(330 - V_s) = 335$

$337.92 - 1.024V_s = 335$

$337.92 - 335 = 1.024V_s$

$2.92 = 1.024V_s$

∴ $V_s = \dfrac{2.92}{1.024} = 2.85$

The student was moving towards his classmate with a velocity of 2.85ms^{-1}

Exercise 10

1. A train approaching a station with a speed of 18ms^{-1} sounds its horn and emits a note of frequency 220Hz. What frequency will be received at the station? (Speed of sound in air is 340ms^{-1}).

2. A blown whistle of frequency 360Hz is moved away from an observer with a speed of 2.5ms^{-1}. What is the apparent frequency of the sound coming to an observer? (Speed of sound in air is 330ms^{-1}).

3. A stationary siren emits a note of frequency 430Hz. What frequency will be received by a train which is approaching it at a speed of 16ms^{-1}? (Speed of sound in air is 340ms^{-1}).

4. A stationary siren at a train station emits a note of frequency 384Hz. What frequency will be received by a train which is leaving the station at a speed of 8ms^{-1}? (Speed of sound in air is 340ms^{-1}).

5. A student sounding a tuning fork of frequency 450Hz moves towards a man with a speed of 3ms^{-1}. If the man is running towards the student with a speed of 4.5ms^{-1}, what is the apparent frequency of the sound coming to the man? (Speed of sound in air is 340ms^{-1}).

6. A football referee and a footballer run away from each other on a football field. The referee runs with a velocity of 1ms^{-1} while the footballer runs with a velocity of 2ms^{-1}. If the referee blows his whistle and emits a note of frequency 350Hz, what is the apparent frequency of the sound that the footballer will hear? (Speed of sound in air is 340ms^{-1}).

7. A football referee and a footballer are running in the same direction with the referee behind the footballer. The referee runs with a velocity of 4ms^{-1} while the footballer runs with a velocity of 2ms^{-1}. If the referee blows his whistle and emits a note of frequency 396Hz, what is the frequency of the sound coming to the footballer? (Speed of sound in air is 340ms^{-1}).

8. A car travelling at a speed of 30ms^{-1} passes a bike which is moving at a speed of 12ms^{-1} in the opposite direction. If the car immediately sounds its horn and emits a note of frequency 464Hz, calculate the apparent frequency that will come to the bike? (Speed of sound in air is 340ms^{-1}).

9. A stationary siren emits a note of frequency 421Hz. If the apparent frequency received by a train which is approaching it at a speed of 10ms^{-1} is 435Hz, determine the speed of sound in air.

10. A student moving towards his classmate blows a whistle and emits a note of frequency 220Hz. If his classmate running towards him at a speed of 4ms^{-1}, hears the sound at an apparent frequency of 225Hz, determine the velocity with which the student was moving towards his classmate. (Speed of sound in air is 340ms^{-1}).

CHAPTER 11
ELECTRIC CURRENT

Electric current is defined as the rate of flow of charge round a circuit. It is denoted by I, and has a unit called ampere, A, which is coulomb/seconds.

Current is given by:
$$I = \frac{Q}{t}$$
where Q is the charge in coulomb, and t is time in seconds.

Potential Difference (P.D)

Potential difference is the work done in joules when one coulomb of charge moves from one point in a circuit to another. It is measured in volt, V.

Resistance

Resistance is defined as the amount of opposition given to the flow of electric current passing through a conductor or circuit. It is denoted by R, and its unit is the Ω. The factors which affect the resistance of a conductor are: length of the conductor, cross-sectional area of the conductor, temperature, and the nature/type of material of the conductor/wire.

Ohm's Law

Ohm's law states that the current flowing through a metallic conductor is directly proportional to the potential difference across its ends, provided that all other physical conditions are constant.

Ohm's law is expressed as follows:
$$V = IR$$
where I is the current and R is a constant called resistance.

Electromotive force (E.M.F)

The e.m.f of a cell is defined as the force needed to drive current through the internal and external resistance of a circuit. E.m.f is expressed as follows:
$$E = I(R + r)$$
where I is the current in the circuit, R is the effective/combined external resistance, and r is the effective/combined internal resistance. Expanding the expression above gives:
$$E = IR + Ir$$
But from Ohm's law, V = IR. Substituting this into the equation above gives:
$$E = V + Ir$$
The expression Ir is known as the lost volt.
$$\text{Lost volt} = Ir$$

Resistivity

The resistivity of a wire is the resistance of a unit cross-sectional area and a unit length of the wire. It is expressed as:

$$p = \frac{RA}{l}$$

where A is the cross-sectional area and L = length of the wire/conductor. The S.I unit of resistivity is Ohm meter (Ωm).

Conductivity

The reciprocal of resistivity gives the conductivity, G, of a wire/conductor. Conductivity is given by:

$$G = \frac{1}{p} \quad \text{or} \quad G = \frac{l}{RA}$$

Examples

1. A 20mC of charge flows through a conductor in 10μsec. Determine the current in the circuit.

<u>Solution</u>
Charge, Q = 20mC, time, t = 10μsec
20mC means 20 milli coulomb, while 10μsec means 10 micro seconds
Hence, 20mC = 20 x 10^{-3} (milli means 10^{-3})
And, 10μsec = 10 x 10^{-6} (micro, μ, means 10^{-6})
Therefore, I = $\frac{Q}{t}$

$$= \frac{20 \times 10^{-3}}{10 \times 10^{-6}}$$
$$= \frac{20}{10} \times (10^{-3--6})$$
$$= 2 \times 10^{-3+6}$$
$$= 2 \times 10^{3} A$$
I = 2000A

2. If the current flowing through a conductor is 4A when the potential difference across it is 6V, calculate the resistance of the conductor.

<u>Solution</u>
From Ohm's law, V = IR
Therefore, R = $\frac{V}{I}$

$$= \frac{6}{4}$$

$$= 1.5\,\Omega$$

3. A nichrome tape of cross-sectional area $5 \times 10^{-6} m^2$ is 5m long. If it has a resistance of 100Ω, calculate:

(a) the resistivity of the tape
(b) the conductivity of the tape

Solutions

(a) Resistivity is given by: $p = \dfrac{RA}{l}$

$$= \frac{100 \times 5 \times 10^{-6}}{5}$$

$$= \frac{5 \times 10^{-4}}{5}$$

$p = 1 \times 10^{-4}\,\Omega m$

(b) Conductivity, $G = \dfrac{1}{p}$

$$= \frac{1}{1 \times 10^{-4}}$$

$= 10^4\,/\Omega m$

$G = 10000\,\Omega m$

4. The quantity of electricity flowing through a wire is found to be 12μC when the current across the wire is 400A. Determine the time taken for the current to flow.

Solution

Quantity of electricity is the charge, Q.
$Q = 20\mu C = 20 \times 10^{-6}$, Current, I = 400A

$$I = \frac{Q}{t} \text{ or } Q = It$$

Therefore, $t = \dfrac{Q}{I}$

$$= \frac{12 \times 10^{-6}}{400}$$

$$= \frac{12 \times 10^{-6}}{4 \times 10^2}$$

$$= \frac{12}{4} \times (10^{-6-2})$$
$$t = 3 \times 10^{-8} \text{sec}$$

5. A voltage of 20V passes through a wire of resistance 8Ω. Calculate the current flowing through the wire.

Solution
From Ohm's law, V = IR

Therefore, $I = \frac{V}{R}$
$$= \frac{20}{8}$$
$$= 2.5 \text{ Amp.}$$

6. A heating element has a cross-sectional area of $6 \times 10^{-4} m^2$. The length of the element is 12m. If it has a resistance of 50Ω, calculate:
(a) the conductivity of the tape
(b) the resistivity of the tape

Solutions

(a) Conductivity is given by: $G = \frac{l}{RA}$
$$= \frac{12}{50 \times 6 \times 10^{-4}}$$
$$= \frac{12}{300 \times 10^{-4}}$$
$$= \frac{12}{3 \times 10^{-2}}$$
$$= 4 \times 10^2 / \Omega m$$

G = 400/Ωm

(b) Resistivity, $p = \frac{1}{G}$
$$= \frac{1}{400}$$
p = 0.0025Ωm
Or, $p = 2.5 \times 10^{-3} \Omega m$

Exercise 11

1. A 10mC of charge flows through a conductor in 200usec. Determine the current in the circuit.

2. If the current flowing through a conductor is 6A when the potential difference across it is 220V, calculate the resistance of the conductor.

3. A nichrome tape of cross-sectional area $2 \times 10^{-6} m^2$ is 50cm long. If it has a resistance of 80Ω, calculate:

(a) the resistivity of the tape

(b) the conductivity of the tape

4. The quantity of electricity flowing through a wire is found to be 0.005C when the current across the wire is 250A. Determine the time taken for the current to flow.

5. A voltage of 10.5V passes through a wire of resistance 3.5Ω. Calculate the current flowing through the wire.

6. A heating element has a cross-sectional area of $10^{-4} m^2$. The length of the element is 1.2m. If it has a resistance of 50Ω, calculate:

(a) the conductivity of the tape

(b) the resistivity of the tape

CHAPTER 12
RESISTORS IN CIRCUITS

Two or more resistors can be connected in series or parallel in a circuit. When such occurs, then there is need to determine the effective/combined resistance of the resistors.

Resistors in Series

Resistors are connected in series when they are connected head to tail in a circuit. In series connection, the same amount of current flows through the resistors, while different amount of voltage passes through them. When n resistors of resistance $R_1, R_2, R_3, ..., R_n$ are connected in series, their effective or total or combined resistance is given by:

$$R = R_1 + R_2 + R_3 + ... + R_n$$

Resistors in Parallel

Resistors are connected in parallel when they are connected end to end in a circuit. In parallel connection, the same amount of voltage flows through the resistors, while different amount of current passes through them. When n resistors of resistance $R_1, R_2, R_3, ..., R_n$ are connected in parallel, their effective or total or combined resistance is given by:

$$\frac{1}{R} = \frac{1}{R_1} + \frac{1}{R_2} + \frac{1}{R_3} + ... + \frac{1}{R_n}$$

The inverse of $\frac{1}{R}$, gives the effective resistance R.

Cells in Circuits

Two or more cells can be connected in series or parallel in a circuit. When cells or battery (two or more cells) are connected together, then there is need to determine the effective/combined e.m.f of the cells.

Cells in Series

When n cells each of e.m.f $E_1, E_2, E_3, ..., E_n$ are connected in parallel, their effective or total or combined emf is given by:

$$E = E_1 + E_2 + E_3 .... + E_n \quad \text{(any one of the emfs of the cells)}$$

The effective internal resistance of the cells is given by:

$$r = r_1 + r_2 + r_3 + ... + r_n$$

Cells in Parallel

When n cells each of e.m.f $E_1, E_2, E_3, ..., E_n$ are connected in parallel, their effective or total or combined emf is given by:

$$E = E_1 \quad \text{(any one of the emfs of the cells)}$$

The effective internal resistance of the cells is given by:

$$\frac{1}{r} = \frac{1}{r_1} + \frac{1}{r_2} + \frac{1}{r_3} + \ldots + \frac{1}{r_n}$$

Therefore, r = the reciprocal of $\frac{1}{r}$

Also, r can be given by:

$r = \frac{r_1}{n}$, where n is the total number of cells.

Examples

1. Two resistors each of resistance 5Ω and 6Ω are connected in series in a circuit. Calculate the effective resistance in the circuit.

<u>Solution</u>

For series combination of resistors, the effective resistance is given by:

$R = R_1 + R_2$

$R = 5 + 6$

$R = 11Ω$

2. Four resistors of 2Ω, 5Ω, 6Ω and 15Ω are connected in parallel across a circuit. Determine the combined resistance of the resistors.

<u>Solution</u>

For parallel combination, the combined resistance is given by:

$$\frac{1}{R} = \frac{1}{R_1} + \frac{1}{R_2} + \frac{1}{R_3} + \frac{1}{R_4}$$

$$= \frac{1}{2} + \frac{1}{5} + \frac{1}{6} + \frac{1}{15}$$

$$= \frac{15 + 6 + 5 + 2}{30}$$

$$= \frac{28}{30}$$

$\frac{1}{R} = \frac{14}{15}$ (In its lowest term)

Therefore, R = $\frac{15}{14}$ (Which is the inverse of $\frac{14}{15}$)

3. Three resistors each of resistance 3Ω are connected in parallel. their combination is then connected in series to a 4Ω resistor, which is also connected in series with a parallel combination of two resistors of 2Ω and 3Ω. Calculate the effective resistance in the circuit.

Solution

The effective resistance of the three parallel resistors each of 3Ω is given by:

$$\frac{1}{R} = \frac{1}{R_1} + \frac{1}{R_2} + \frac{1}{R_3}$$

$$= \frac{1}{3} + \frac{1}{3} + \frac{1}{3}$$

$$= \frac{1+1+1}{3}$$

$$= \frac{3}{3}$$

$$\frac{1}{R} = 1$$

Therefore, R = 1 (Which is the inverse of 1)

The effective resistance of the two parallel resistors of 2Ω and 3Ω is given by:

$$\frac{1}{R} = \frac{1}{R_1} + \frac{1}{R_2}$$

$$= \frac{1}{2} + \frac{1}{3}$$

$$= \frac{3+2}{6}$$

$$\frac{1}{R} = \frac{5}{6}$$

R = $\frac{6}{5}$ (Which is the inverse of $\frac{5}{6}$)

R = 1.2

Therefore, the combinations have now been reduced to series values connections, of 1Ω (from the three parallel combination of 3Ω each), the 4Ω at the middle, and 1.2 (from the parallel combination of 2Ω and 3Ω resistors)

These three series combination now gives

R = 1 + 4 + 1.2

R = 6.2Ω

4. Three cells each of emf 2V and internal resistance 1Ω are connected in series. Calculate the effective emf and internal resistance of the cells.

Solution

Effective emf for a series combination of cells is given by:

E = $E_1 + E_2 + E_3$

= 2 + 2 + 2

E = 6V

The effective internal resistance for a series combination of cells is given by:

r = $r_1 + r_2 + r_3$

$$= 1 + 1 + 1$$
$$r = 3\Omega$$

5. Four cells each of emf 1.5V and internal resistance 2Ω are connected in parallel. Determine the combined emf and internal resistance of the cells.

Solution

The combined emf for a parallel combination of cells is given by:

$E = E_1$ (Which is the emf of one of the cells)

$E = 1.5V$

The effective internal resistance for a parallel combination of cells is given by:

$$\frac{1}{r} = \frac{1}{r_1} + \frac{1}{r_2} + \frac{1}{r_3} + \frac{1}{r_4}$$

$$= \frac{1}{2} + \frac{1}{2} + \frac{1}{2} + \frac{1}{2}$$

$$\frac{1}{r} = \frac{1+1+1+1}{2}$$

$$= \frac{4}{2}$$

$$\frac{1}{r} = 2$$

Therefore, $r = \frac{1}{2}$ (Which is the inverse of 2)

Or, $r = 0.5\Omega$

Alternatively, r can be obtained by using: $r = \frac{r_1}{n}$, where r_1 is one of the internal resistances, while n is the number of cells.

$$r = \frac{r_1}{n}$$

$$= \frac{2}{4}$$

$r = 0.5\Omega$ (As obtained before)

Exercise 12

1. Two resistors each of resistance 3.2Ω and 5.5Ω are connected in series in a circuit. Calculate the effective resistance in the circuit.

2. Four resistors of 1Ω, 2Ω, 3Ω and 4Ω are connected in parallel across a circuit. Determine the combined resistance of the resistors.

3. Three resistors each of resistance 4Ω are connected in parallel. Their combination is then connected in series to a 2Ω resistor, which is also connected in series with a parallel

combination of two resistors of 5Ω and 10Ω. Calculate the effective resistance in the circuit.

4. Three cells each of emf 1.5V and internal resistance 1.2Ω are connected in series. Calculate the effective emf and internal resistance of the cells.

5. Four cells each of emf 2V and internal resistance 5Ω are connected in parallel. Determine the combined emf and internal resistance of the cells.

CHAPTER 13
DIVISION OF CURRENT AND VOLTAGES BETWEEN RESISTORS IN CIRCUITS

Recall that when resistors are connected in series, the same amount of current flows through the resistors, while different amount of voltage passes through them. In parallel connection, the same amount of voltage flows through the resistors, while different amount of current passes through them. These principles can be used to determine the current and voltages passing through resistors connected in electrical circuits.

Examples

1. A current of 1.5A flows in a circuit containing two resistors each of 2Ω and 3Ω which are connected in series. Determine:
(a) the current flowing through each resistor
(b) the voltage across each resistor
(c) the total voltage in the circuit

Solution
(a) Since they are connected in series, the same amount of current will flow through them. Hence, 1.5A will flow through the 2Ω resistor, and the same 1.5A will flow through the 3Ω resistor.

(b) The voltage across each resistor can be determine by using the expression for Ohm's law.
Therefore, for the 2Ω resistor, V = IR
$\quad\quad\quad$ = 1.5 x 2
$\quad\quad$ V = 3V
For the 3Ω resistor, V = IR
$\quad\quad\quad$ = 1.5 x 3
$\quad\quad$ V = 4.5V

(c) The total voltage in the circuit is obtained by a combination of the voltage across each resistor.
$\quad$Total voltage = 3 + 4.5

$\quad\quad\quad\quad$ = 7.5V

2. A voltage of 20V in a circuit supplies current to three resistors each of 1Ω, 2Ω and 5Ω which are connected in series. Determine:
(a) the current flowing through each resistor
(b) the voltage across each resistor

Solution
(a) Let us first find the effective resistance so that we can find the current in the circuit.
Effective resistance, R = 1 + 2 + 5 = 8Ω
Therefore the current in the circuit is given by:

$$V = IR$$
Or, $I = \dfrac{V}{R}$
$= \dfrac{20}{8}$
I = 2.5A

Since the resistors are all connected in series, the same current of 2.5A will flow through each of them

(b) Different voltages will pass across them.
Therefore, for the 1Ω resistor: V = IR
= 2.5 x 1
V = 2.5V

For the 2Ω resistor: V = IR
= 2.5 x 2
V = 5V

For the 5Ω resistor: V = IR
= 2.5 x 5
V = 12.5V

Note that adding all the voltages across all the resistors will give the total voltage of 20V (2.5 + 5 + 12.5 = 20) in the circuit.

3. A current of 2A flows to the junction of two resistors each of 3Ω and 6Ω which are connected in parallel in a circuit. Determine:
(a) the current flowing through each resistor
(b) the voltage across each resistor
(c) the total voltage in the circuit

Solution
(a) Since they are in parallel, different current will pass through them. The combined resistance is give by:

$$\frac{1}{R} = \frac{1}{R_1} + \frac{1}{R_2}$$
$$= \frac{1}{3} + \frac{1}{6}$$
$$= \frac{2+1}{6}$$
$$\frac{1}{R} = \frac{3}{6}$$
$$R = \frac{6}{3}$$
$$R = 2$$

Therefore, from Ohm's law, V = IR
$$= 2 \times 2$$
$$= 4 \text{ V}$$

This voltage of 4V which is the voltage in the circuit, will flow across each of the resistor since they are in parallel.

Therefore the current across each resistor can be obtained by:
$$I = \frac{V}{R} \quad \text{(From V = IR)}$$

For the 3Ω resistor, $I = \frac{V}{R}$
$$= \frac{4}{3} \text{ A}$$

For the 6Ω resistor, $I = \frac{V}{R}$
$$= \frac{4}{6}$$
$$= \frac{2}{3} \text{ A}$$

Note that the total current of 2A = $\frac{4}{3} + \frac{2}{3}$ (which is the sum of the current across each resistor)

Alternatively, this question can be solved as follows:

Since the resistors are 3Ω and 6Ω, then the current getting to the junction will be shared between the two resistors in the ratio 6 : 3 respectively. This can be arranged as follows:

Resistors: 3 and 6
Ratios of current: 6 : 3   (Just simply reverse the resistors)
Ratios of current: 2 : 1   (In its lowest term)

Therefore, the 2A current is now share as follows:
Total ratio = 2 + 1 = 3

Hence current across 3Ω resistor (having ratio value of 2) is: $\frac{2}{3} \times 2 = \frac{4}{3}$ A

Current across 6Ω resistor (having ratio value of 1) is: $\frac{1}{3} \times 2 = \frac{2}{3}$ A

These values are as obtained before.

Note that the higher the resistor, the lower the current, and the lower the resistor, the higher the current.

Note also that this second method can only be used when two resistors are connected in parallel.

(b) The voltage across each resistor is obtained from: V = IR

Therefore for the 3Ω resistor: V = IR

$$= \frac{4}{3} \times 3$$

$$= 4V$$

For the 6Ω resistor: V = IR

$$= \frac{2}{3} \times 6$$

$$= 4V$$

Note that the two voltages have to be equal since the resistors are in parallel.

(c) The total voltage in the circuit is 4V. This is the voltage across anyone of the resistors since they are in parallel.

4. A voltage of 10V flows to the junction of three resistors each of 2Ω, 3Ω and 6Ω which are connected in parallel. Determine:

(a) the current flowing through each resistor
(b) the voltage across each resistor
(c) the total current in the circuit

Solution

(a) Since the resistors are connected in parallel, then the same voltage of 10V will pass across each of them.

Therefore the current through each of them is obtained as follows:

Current across the 2Ω resistor is: $I = \frac{V}{R}$

$$= \frac{10}{2} = 5A$$

Current across the 3Ω resistor is: $I = \frac{V}{R}$

$$= \frac{10}{3}$$

Current across the 6Ω resistor is: $I = \frac{V}{R}$

$$= \frac{10}{6}$$

$$= \frac{5}{3}A$$

(b) The voltage across each resistor will be 10V since the resistors are in parallel

(c) The total current in the circuit can be obtained by adding the current across each of the resistors. This gives:

$$I = 5 + \frac{10}{3} + \frac{5}{3}$$

$$I = 10A$$

Alternatively, the effective resistance can be obtained first. This gives:

$$\frac{1}{R} = \frac{1}{2} + \frac{1}{3} + \frac{1}{6}$$

$$= \frac{3+2+1}{6}$$

$$\frac{1}{R} = \frac{6}{6} = 1$$

$$R = 1$$

From Ohm's law, V = IR

Therefore, $I = \frac{V}{R}$

$$= \frac{10}{1}$$

$$= 10A$$

5. Three resistors each of resistance 2Ω are connected in parallel. Their combination is then connected in series to a 1Ω resistor, which is also connected in series with a parallel combination of two resistors of 3Ω and 9Ω. If a current of 0.6A flows through the 1Ω resistor, calculate:

(a) the current in the circuit
(b) the current across each resistor
(c) the voltage across each resistor
(d) the voltage in the circuit

Solution

(a) The 1Ω resistor is alone in the circuit, and it is connected in series with the combination of the other two sets of parallel combination. Since the current in the 1Ω resistor is 0.6A, it also means that the current that gets to each of the junction of the resistors connected in parallel is 0.6A. This is because the same current flows through resistors in series, and it is this same current that flows in the circuit.

Hence the current in the circuit is 0.6A.

(b) The current of 0.6A flows to the junction of the three 2Ω resistors connected in parallel. Different current will pass through them since they are in parallel. Since the three 2Ω resistors are equal in resistance, then the current will be shared equally among them. Therefore, the current will be shared into three as follows: $\frac{0.6}{3} = 0.2$

Hence the current across each of the three 2Ω resistors is 0.2A

Similarly, the current to the junction of the 3Ω and 9Ω resistors in parallel is 0.6A. This current will be shared as follows:

Resistors: 3 and 9

Ratios of current: 9 : 3   (Just simply reverse the resistors values)

Ratios of current: 3 : 1   (In its lowest term)

Therefore, the 0.6A current is now shared as follows:

Total ratio = 3 + 1 = 4

Hence current across 3Ω resistor (having ratio value of 3) is: $\frac{3}{4}$ x 0.6 = 0.45A

Current across 9Ω resistor (having ratio value of 1) is: $\frac{1}{4}$ x 0.6 = 0.15A

The current across each resistor can now be stated as follows:

Current across each of the 2Ω resistors in parallel = 0.2A

Current across the 1Ω resistor = 0.6A   (As stated in the question)

Current across the 3Ω resistor = 0.45A

Current across the 9Ω resistor = 0.15A

(c) The voltage across each of the resistor is calculated from V = IR as follows:

Current across each of the 2Ω resistors in parallel = 0.2A, therefore voltage across each of them is given by: V = IR

= 0.2 x 2

= 0.4V

Current across the 1Ω resistor = 0.6A, therefore voltage across it is given by: V = IR

= 0.6 x 1

= 0.6V

Current across the 3Ω resistor = 0.45A, therefore voltage across it is given by: V = IR

= 0.45 x 3

= 1.35V

Voltage across the 9Ω resistor is also 1.35V, (i.e. 0.15 x 9 = 1.35) since it is in parallel with the 3Ω resistor.

(c) The voltage in the circuit can be obtained by adding the voltage from each of the group of resistors in the circuit.

The voltage from either of the three 2Ω resistor is 0.4V

The voltage from the 1Ω resistor is 0.6V

The voltage from either of the 3Ω or 9Ω resistor is 1.35V

Therefore, the voltage in the circuit is: 0.4 + 0.6 + 1.35 = 2.35V

6. A current of 4A flows in a circuit containing two resistors each of 3Ω and RΩ which are connected in series. If the voltage in the circuit is 18V, determine:

(a) the voltage across the RΩ resistor

(b) the value of the resistance R.

Solution

(a) The voltage across the 3Ω resistor is given by: V = IR

$$= 4 \times 3 = 12V$$

Therefore the remaining voltage in the circuit will pass through the RΩ resistor. This is given by:

$$18 - 12 = 6V$$

(b) Since the two resistors are in series, the same current will flow across them.

Therefore, current across the RΩ resistor is 4A, while the voltage across it, is 6V.

From Ohm's law: V = IR

So, for the RΩ resistor: V = IR

$$6 = 4 \times R$$

$$\text{Therefore, } R = \frac{6}{4}$$

$$R = 1.5\Omega$$

7. A current of 9A flows to the junction of three resistors each of 1Ω, RΩ and 4Ω which are connected in parallel in a circuit. If the voltage across the 4Ω resistor is 6V, Determine:

(a) the current flowing through each resistor

(b) the value of R.

Solution

(a) Since the resistors are connected in parallel, then the same voltage of 6V passes through each of them. Hence the current across each of them can be obtained by: V = IR, so that: $I = \frac{V}{R}$

Therefore current flowing through the 1Ω resistor is given by:

$$I = \frac{V}{R}$$

$$= \frac{6}{1}$$

$$I = 6A$$

Current flowing through the 4Ω resistor is given by:

$$I = \frac{V}{R}$$

$$= \frac{6}{4}$$

$$I = 1.5A$$

Current flowing through the RΩ resistor is given by:

$I = 9 - (6 + 1.5)$ (Since total current in the circuit is 9A)
$= 9 - 7.5$
$I = 1.5A$

(b) The voltage across the RΩ resistor is 6V. The current across the RΩ resistor is 1.5V. Therefore the value of R is obtained as follows:

$V = IR$
$6 = 1.5 \times R$
$R = \dfrac{6}{1.5}$
$R = 4A$

Note that when resistors ire in series, the sum of the voltages across each of them gives the total voltage in the circuit, but the current across each of them cannot be added to give the total current in the circuit. The same current flows across each of them, and that is the current in the circuit.

Also, when resistors are in parallel, the sum of the current across each of them gives the total current in the circuit, but the voltage across each of them cannot be added to give the total voltage in the circuit. The same voltage flows across each of them, and that is the voltage in the circuit.

Exercise 13

1. A current of 1A flows in a circuit containing two resistors each of 5Ω and 9Ω which are connected in series. Determine:
(a) the current flowing through each resistor
(b) the voltage across each resistor
(c) the total voltage in the circuit

2. A voltage of 12V in a circuit supplies current to three resistors each of 2Ω, 4Ω and 5Ω which are connected in series. Determine:
(a) the current flowing through each resistor
(b) the voltage across each resistor

3. A current of 3.5A flows to the junction of two resistors each of 1Ω and 3Ω which are connected in parallel in a circuit. Determine:
(a) the current flowing through each resistor
(b) the voltage across each resistor
(c) the total voltage in the circuit

4. A 50V supply flows to the junction of three resistors each of 5Ω, 8Ω and 10Ω which are connected in parallel. Determine:
(a) the current flowing through each resistor

(b) the voltage across each resistor
(c) the total current in the circuit

5. Three resistors each of resistance 6Ω are connected in parallel. Their combination is then connected in series to a 5Ω resistor, which is also connected in series with a parallel combination of two resistors of 1Ω and 4Ω. If a current of 1.8A flows through the 5Ω resistor, calculate:
(a) the current in the circuit
(b) the current across each resistor
(c) the voltage across each resistor

6. A current of 2.2A flows in a circuit containing two resistors each of 20Ω and XΩ which are connected in series. If the voltage in the circuit is 55V, determine:
(a) the voltage across the XΩ resistor
(b) the value of the resistance X.

7. A current of 8.75A flows to the junction of three resistors each of 8Ω, RΩ and 2Ω which are connected in parallel in a circuit. If the voltage across the 8Ω resistor is 10V, Determine:
(a) the current flowing through each resistor
(b) the value of R.

CHAPTER 14
GENERAL CALCULATIONS IN ELECTRIC CIRCUITS

Examples

1. Two cells, each of emf 1.2V and internal resistance 2Ω are connected in series. A 4Ω resistor is connected in series with the cell. Calculate:
(a) the current flowing in the circuit
(b) the lost volt.

<u>Solution</u>
(a) Effective emf, E = 1.2 + 1.2   (when cells are in series, add the emf of each of the cells to get the effective emf)
$$E = 2.4V$$
Effective internal resistance, r = 2 + 2   (when cells are in series, add the internal resistance of each of the cells to get the effective internal resistance)
$$r = 4Ω$$
Effective external resistance, R = 4
Therefore, in order to obtain the current, we use the emf expression given by:
$$E = I(R + r)$$
$$2.4 = I(4 + 4)$$
$$2.4 = 8I$$
$$I = \frac{2.4}{8}$$
$$I = 0.3A$$
(b) Lost volt = Ir
$$= 0.3 \times 4$$
$$= 1.2V$$

2. Two resistors, 2Ω and 5Ω are joined in parallel and their terminals are connected to the poles of a cell. A current of 3A flows in the 2Ω resistor. What is the total current passed by the cell?

<u>Solution</u>
The voltage across the 2Ω resistor is given by:
$$V = IR$$
$$= 3 \times 2$$
$$= 6V$$
This same amount of voltage will also pass through the 5Ω resistor since the two resistors are connected in parallel. Hence the current across the 5Ω resistor is obtained as follows:
$$V = IR$$

Therefore, $I = \dfrac{V}{R}$

$= \dfrac{6}{5}$

$= 1.2A$

Now, the total current in the circuit is the sum of the current across the two resistors.

$I = 3 + 1.2$

$I = 4.2A$

Therefore the total current passed by the cell is 4.2A.

3. A battery consists of two cells joined in parallel and each has an emf of 1.5V and an internal resistance of 4Ω.

(a) What current will flow in the circuit if two resistors of 2Ω each are connected in series across the circuit?

(b) What is the lost volt in the circuit?

Solution

(a) Effective emf, E = 1.5V (when cells are in parallel, take the emf of just one of the cells to get the effective emf)

Effective internal resistance, $r = \dfrac{r_1}{n}$ (when cells are in parallel, take one of the values of the internal resistance and divide it by the number of cells to get the effective internal resistance)

$r = \dfrac{4}{2}$

$r = 2Ω$

Effective external resistance, R = 2 + 2 (Resistors in series are added to get effective resistance)

R = 4Ω

Therefore, in order to obtain the current, we use the emf expression given by:

E = I(R + r)

1.5 = I(4 + 2)

1.5 = 6I

$I = \dfrac{1.5}{6}$

I = 0.25A

(b) The lost volt is given by:

Lost volt = Ir (r is the effective internal resistance)

= 0.25 x 2

= 0.5V

4. A cell of emf 2V is in series with a resistor of 12Ω. A high resistance voltmeter put across the cell reads 1.2V. What is the internal resistance of the cell?

Solution

The potential difference p.d, is given by:

$V = IR$

$2 = I \times 12$ (The resistance is 12Ω, while the voltmeter is 2V)

Therefore, $I = \dfrac{2}{12}$

$= \dfrac{1}{6}$

$I = 0.167A$

From the emf formula:

$E = I(R + r)$

Or, $E = IR + Ir$

Or, $E = V + Ir$ (Since $V = IR$)

$2 = 1.2 + Ir$ (Since $V = 1.2V$)

$Ir = 2 - 1.2$

$Ir = 0.8$

Therefore, $r = \dfrac{0.8}{I}$

$= \dfrac{0.8}{0.167}$

$r = 1.05V$

5. A battery consist of three cells joined in parallel and each has an emf of 2V and an internal resistance of 1.5Ω. What current will flow in the circuit if two resistance of 8Ω each are connected in parallel across the circuit? Determine the lost voltage in the circuit.

Solution.

Effective emf, $E = 2V$ (when cells are in parallel, take the emf of just one of the cells to get the effective emf)

Effective internal resistance, $r = \dfrac{r_1}{n}$ (when cells are in parallel, take one of the values of the internal resistance and divide it by the number of cells to get the effective internal resistance)

$r = \dfrac{1.5}{3}$

$r = 0.5Ω$

Effective external resistance, $\dfrac{1}{R} = \dfrac{1}{8} + \dfrac{1}{8}$ (For resistors in parallel, first add the reciprocal of their values to get the inverse of the effective resistance)

$\dfrac{1}{R} = \dfrac{2}{8}Ω$

$$R = \frac{8}{2}$$
$$R = 4\Omega$$

Alternatively, since the external resistance are equal, the effective value can be obtained by: $R = \frac{R_1}{n}$ (In a similar way to internal resistance in parallel)

Therefore, $R = \frac{8}{2}$

$R = 4\Omega$ as obtained before.

Therefore, in order to obtain the current, we use the emf expression given by:

$E = I(R + r)$
$2 = I(4 + 0.5)$
$2 = 4.5I$
$I = \frac{2}{4.5}$
$I = 0.44A$

6. A series combination of two resistors of resistances RΩ and 4Ω is connected to the terminals of a battery of emf 20V and negligible internal resistance. If the current in the circuit is 3A, determine the value of R.

Solution

Since the internal resistance is negligible, we apply Ohm's law as follows:

$V = IR$

Or, $E = IR$ (No internal resistance)

The effective external resistance is given by: R + 4 (we add them since they are in series)

Therefore, $E = IR$

$20 = 3 \times (R + 4)$
$20 = 3R + 12$ (After expanding the bracket)
$20 - 12 = 3R$
$8 = 3R$
$R = \frac{8}{3}\Omega$

Or, $R = 2.67\Omega$

Exercise 14

1. Two cells, each of emf 1.4V and internal resistance 0.5Ω are connected in series. A 10Ω resistor is connected in series with the cell. Calculate:

(a) the current flowing in the circuit

(b) the lost volt

2. Two resistors, 5Ω and 8Ω are joined in parallel and their terminals are connected to the poles of a cell. A current of 2.5A flows in the 8Ω resistor. What is the total current passed by the cell?

3. A battery consists of two cells joined in parallel and each has an emf of 1.25V and an internal resistance of 0.2Ω.

(a) What current will flow in the circuit if two resistors of 5Ω each are connected in series across the circuit?

(b) What is the lost volt in the circuit?

4. A cell of emf 1.5V is in series with a resistor of 3Ω. A high resistance voltmeter put across the cell reads 1.2V. What is the internal resistance of the cell?

5. A battery consists of five cells joined in parallel and each has an emf of 1.08V and an internal resistance of 10Ω. What current will flow in the circuit if two resistors of 2Ω each are connected in parallel across the circuit? Determine the lost voltage in the circuit.

6. A parallel combination of two resistors of resistances YΩ and 2.5Ω is connected to the terminals of a battery of emf 10V and negligible internal resistance. If the current in the circuit is 8A, determine the value of Y.

CHAPTER 15
ELECTRICAL ENERGY

Electrical energy is simply defined as the work done in a circuit. Therefore if the pd applied is V volts and the quantity of electricity is Q coulombs, the work done is given by:

$$W = VQ$$

Or, $W = VIt$ (Since $Q = It$)

Therefore, electrical energy is given by:

$$E = IVt \quad \text{(The time should be in seconds)}$$

By applying Ohm's law, electrical energy can also be expressed as:

$$E = I^2Rt \quad \text{(Since } V = IR\text{)}$$

Or, $E = \dfrac{V^2 t}{R}$ (Since $I = \dfrac{V}{R}$)

Electrical Power

Electrical power is the rate at which electrical energy is expended. It is expressed as:

$$\text{Power} = \dfrac{\text{Electrical Energy}}{\text{time}}$$

That is: $P = \dfrac{E}{t}$

Therefore we can write power from the three formulas for electrical energy above as follows:

$$P = IV$$

Or, $P = I^2R$

Or, $P = \dfrac{V^2}{R}$

These are obtained when all the formulas for electrical energy are each divided by time.

The unit of power is watt (W). The larger unit is the kilowatt, KW.

$$1KW = 1000W$$

To convert from watts to kilowatts, we simply divide by 1000. For example, 2500W = $\left(\dfrac{2500}{1000}\right)KW = 2.5KW$

Examples

1. How much heat energy is produced in two minutes by an electrical iron which draws 6.0A when connected to a 220V supply?

Solution

From the question we can see that: Current, I = 6A, voltage, V = 220V, t = 2 x 60 = 120sec (multiply number of minutes by 60 in order to convert it to seconds)

Therefore, heat produced = Electrical energy, which is given by:

E = IVt   (There are three formulas for electrical energy, but this is the one that relates current, voltage and time)

E = 6 x 220 x 120

= 158400J

E = 158.4KJ   (Divide by 1000 to convert to KJ)

2. A filament lamp is rated 220V, 40W. What does this mean? Calculate:
(a) the current which is drawn from the lamp
(b) the resistance of the filament

Solution

The rating 220V, 40W means that at a voltage supply of 220V, the lamp will draw up a power of 40W for its operation.

(a) V = 220V, P = 40W, while I = ?

The electrical power formula which connects voltage, V and current, I is given by:

Power = IV

40 = 220I

Therefore, I = $\frac{40}{220}$

I = 0.18A

(b) V = 220V, P = 40W, R = ?

The electrical power formula which connects voltage, V and resistance, R is given by:

$P = \frac{V^2}{R}$

$40 = \frac{220^2}{R}$

$R = \frac{220^2}{40}$

$= \frac{220 \times 220}{40}$

R = 1210Ω

3. A radio takes a current of 0.6A from a 15V supply. Calculate the energy produced in 10 minutes.

Solution

E = IVt

= 0.6 x 15 x (10 x 60)   (Note that 10 minutes = 10 x 60 seconds)

= 5400J

4. A current of 0.5A is drawn by an electrical device of resistance 420Ω. If this device is operated for 2 hours, calculate:

(a) the power of the device

(b) the energy expended.

Solution

(a) I = 0.5A, R = 420Ω, P = ?

The electrical power formula which connects current, V and resistance, R is given by:

$P = I^2R$

$= 0.5^2 \times 420$

$= 0.25 \times 420$

P = 105W

(b) 2 hours = (2 x 60 x 60)seconds   (In order to convert hours to seconds, simply multiply the time by 60 x 60. The first 60, converts hours to minutes, while the second 60 converts minutes to seconds)

Hence, 2 hours = 2 x 60 x 60 = 7200sec.

Recall that: Power = $\frac{\text{Electrical energy}}{\text{time}}$

That is: $P = \frac{E}{t}$

Therefore,   E = Pt

= 105 x 7200

E = 756000J

E = 756KJ

5. The heat produced by a refrigerator when operated for 1 day was found to be 80,000KJ. If the voltage supply is 210V, calculate:

(a) the current drawn by the refrigerator

(b) the resistance of the heating material of the refrigerator

Solution

(a) Heat energy, E = 80,000KJ = 80,000,000J (Multiply by 1000 to convert KJ to J)

Voltage, V = 210V, time, t = 1 day = 24 hours = (24 x 60 x 60)seconds

Therefore, E = IVt

80,000,000 = I x 210 x (24 x 60 x 60)

80,000,000 = 18,144,000I

Therefore, I = $\frac{80000000}{18144000}$

I = 4.41A

(b) $E = \dfrac{V^2 t}{R}$

$R = \dfrac{V^2 t}{E}$

$= \dfrac{220^2 \times (24 \times 60 \times 60)}{80000000}$

$R = 47.6 \Omega$

6. An alternating current with an rms value of 5A passes through a resistor of 20Ω. Determine:
(a) the power through the resistor
(b) the energy that will run through the resistor for 38sec.

Solution
(a) The formula for power when current, I, and resistance, R, are given is:

$P = I^2 R$
$= 5^2 \times 20$
$= 25 \times 20$
$= 500W$

(b) Energy = Power x time
$E = Pt$
$= 500 \times 38$
$E = 19000J$
Or, $E = 19KJ$

Exercise 15

1. How much heat energy is produced in 5 minutes by an electrical iron which draws 12.0A when connected to a 240V supply?
2. A filament lamp is rated 240V, 60W. What does this mean? Calculate:
(a) the current which is drawn from the lamp
(b) the resistance of the filament
3. A radio takes a current of 1.2A from a 10V supply. Calculate the energy produced in 1 hour.
4. A current of 2.1A is drawn by an electrical device of resistance 10Ω. If this device is operated for 1 day, calculate:
(a) the power of the device
(b) the energy expended.

5. The heat produced by a refrigerator when operated for 4 hours was found to be 200KJ. If the voltage supply is 220V, calculate:
(a) the current drawn by the refrigerator
(b) the resistance of the heating material of the refrigerator

6. An alternating current with an rms value of 8A passes through a resistor of 960Ω. Determine:
(a) the power through the resistor
(b) the energy that will run through the resistor for 12sec.

CHAPTER 16
BUYING OF ELECTRICAL ENERGY

Electrical energy is usually sold to the public in KWh (kilowatt hour) units. 1KWh is the amount of energy supplied when 1000W works for one hour.
This means that:
    1KWh = 1000W for 1 hour
In order to carry out calculations on buying of electricity, it is important to first calculate the power, then convert the power in watt to power in kilowatt. After which you multiply the power in kilowatt by time in hours. This will give you the energy used up in KWh.

Examples
1. An electric heater takes up 5A when operated on a 240V supply. Calculate the cost of electricity consumed at 9 cents per KWh when the heater is used for 6 hours.
Solution
Power, P = IV
    = 5 x 240
    P = 1200W
Power in KW = 1200/1000
    = 1.2KW
Energy in KWh = 1.2 x 6   (That is, power in KW multiplied by time in hours)
    E = 7.2KWh
Now, the cost of electricity from the question is 9 cents per KWh
Therefore, 7.2KWh will cost:
    7.2 x 9
    = 648 cents
Cost of electricity in dollars = $\frac{648}{100}$ (Divide cost in cents by 100 in order to convert the cost to dollars)
    = $6.48

2. An electric heater (1.2KW), a fan (520W), a water boiling ring (1100W) and 12 bulbs (40W each) are used in a home. How much will it cost to run these appliances for one month if the cost of electricity is 10.5cents per KWh?
Solution
The powers of these appliances have already been given. Let us convert them to power in KW.
    Electric heater = 1.2KW

Fan = 520W = $(\frac{520}{1000})$KW = 0.52KW

Boiling ring = 1100W = $(\frac{1100}{1000})$KW = 1.1KW

12 Bulbs = 40 x 12 = 480W = $(\frac{480}{1000})$KW = 0.48 KW

Therefore, total power in KW = 1.2 + 0.52 + 1.1 + 0.48
= 3.3KW

Let us express the time in hours as follows:

1 month = 30 days = (30 x 24)hours = 720hours

Therefore, Energy in KWh = Power in KW x time in hours
= 3.3 x 720
= 2376KWh

Now, 1KWh of electricity cost 10.5 cents (As given in the question)

Therefore, 2376KWh will cost:

2376 x 10.5
= 24948 cents

Cost in dollars = $\frac{24948}{100}$
= $249.48

3. An electric pump rated 2hp, 240V, is used to pump water for 30 minutes. If it operates normally and the cost of electricity is 12 cents per KWh, how much does it cost to pump water for this period?

Solution

Hp means horse power.

1hp = 746W

Therefore, 2hp = 746 x 2 = 1492W

Power in KW = $\frac{1492}{1000}$
= 1.492KW

Time in hours is given by:

30 minutes = $(\frac{30}{60})$hours   (Divide time in minutes by 60 in order to convert it to time in hours)

= 0.5hours

Therefore, Energy in KWh = 1.492 x 0.5
= 0.746KWh

Since 1KWh of electricity cost 12 cents (from question), then:

0.746KWh will cost:

0.746 x 12
= 8.952 cents

= 9 cents   (Aproximately)

4. An electric iron (1000W), 2 televisions (80W each), a refrigerator (900W), an electric heater (1.2KW), and 6 bulbs (60W each) are used in a house. What amount will it cost to put on all the appliances for 7 hours if 1KWh of electricity cost 10 cents.
Solution
The powers of these appliances have already been given. Let us convert them to power in KW.

Electric iron = 1000W = $(\frac{1000}{1000})$KW = 1KW

2 televisions = 2 x 80 = 160W = $(\frac{160}{1000})$KW = 0.16KW

Refrigerator = 900W = $(\frac{900}{1000})$KW = 0.9KW

Electric heater = 1.2KW

6 Bulbs = 60 x 6 = 360W = $(\frac{360}{1000})$KW = 0.36 KW

Therefore, total power in KW = 1 + 0.16 + 0.9 + 1.2 + 0.36
= 3.62KW

Time in hours = 7 hours
Therefore, Energy in KWh = Power in KW x time in hours
= 3.62 x 7
= 25.34KWh

Now, 1KWh of electricity cost 10cents (As given in the question)
Therefore, 25.34KWh will cost:
25.34 x 10
= 253.4 cents

Cost in dollars = $\frac{253.4}{100}$
= $2.53

5. An electric heater takes up 6A when operated on a 220V supply, while an air conditioner takes up 5A. If they are both run for 12 hours, how much will it cost, at 9cents per KWh?
Solution
Power of the electric heater, P = IV
= 6 x 220
= 1320W

Therefore power in KW = $\frac{1320}{1000}$ = 1.32KW

Power of the air conditioner, P = IV
= 5 x 220

= 1100W

Therefore power in KW = $\frac{1100}{1000}$ = 1.1KW

Total power of the appliances = 1.32 + 1.1

= 2.42KW

Hence, Energy in KWh = 2.42 x 12

= 29.04KWh

Now, 1KWh cost 9 cents

Therefore, 29.04KWh will cost:

29.04 x 9

= 261.36cents

Cost in dollars = $\frac{261.36}{100}$

= $2.61

Exercise 16

1. An electric heater takes up 2.4A when operated on a 110V supply. Calculate the cost of electricity consumed at 11cents per KWh when the heater is used for 1 day.

2. An electric heater (1KW), a fan (320W), a water boiling ring (1200W) and 8 bulbs (20W each) are used in a home. How much will it cost to run these appliances for one week if the cost of electricity is 10cents per KWh?

3. An electric pump rated 1.5KW, 220V, is used to pump water for 18 minutes. If it operates normally and the cost of electricity is 9 cents per KWh, how much does it cost to pump water for this period?

4. An electric heater (900W), 2 televisions (120W each), a refrigerator (600W), an electric iron (1.1KW), and 10 bulbs (40W each) are used in a house. What amount will it cost to put on all the appliances for 2 hours if electricity cost 10.2 cents for 1KWh.

5. An electric heater takes up 10A when operated on a 450V supply, while an air conditioner takes up 8A. If they are both run for 18 hours, how much will it cost, at 9.5cents per KWh?

CHAPTER 17
MEASUREMENT OF RESISTANCE

Wheatstone Bridge Method
In a Wheatstone bride, two resistors, R_1 and R_2 are connected in series in a circuit. Another two resistors R_3 and R_4 are also connected in series. These pairs of resistors in series are then connected in parallel. A centre-zero galvanometer is connected to the middle of the parallel combination. One of the resistors will be unknown. By adjusting the values of the resistors, the current in the centre zero galvanometer can be made zero. At this balance point, the following expression is valid.
$$\frac{R_1}{R_2} = \frac{R_3}{R_4}$$

Metre Bridge Method
The metre bridge is a form of a wheatstone bridge in which two of the resistors are replaced by a uniform wire of 1 metre. When the bridge is balanced, the expression below follows.
$$\frac{R_1}{R_2} = \frac{l_1}{l_2}$$
Where L_1 is the balance point from the left hand side of the metre bridge. Since the wire in the metre bridge is 1 metre (100cm) long, then L_2, is 100 - L_1

Comparing e.m.f of cells by using potentiometer
A potentiometer can be used to compare the emfs of two cells. The expression used for this comparison is given by:
$$\frac{E_1}{E_2} = \frac{l_1}{l_2}$$
However, in the case of a potentiometer, circuit where the resistance of the potentiometer wire is given, and other resistors are also connected to the circuit, the emf to be determined is related to the voltage across the potentiometer wire as follows:
$$\frac{E}{V} = \frac{l}{100}$$
where E is the emf to be determined, V is the voltage across the potentiometer wire, and L is the balance length. 100 represent the length of the potenticmeter wire which is usually 100cm.

Examples
1. The balance length on a metre bridge for two resistors, A and B is 40cm on the side of A. If A is 12Ω, calculate the value of B.
<u>Solution</u>
For a metre bridge at balance point:

$$\frac{R_1}{R_2} = \frac{l_1}{l_2}$$

By using A for R_1 and B for R_2, the expression above becomes:

$$\frac{A}{B} = \frac{l_1}{l_2}$$

$$\frac{12}{B} = \frac{40}{100 - 40} \quad \text{(Note that } L_2 = 100 - L_1 \text{ since the wire is 1m long)}$$

$$\frac{12}{B} = \frac{40}{60}$$

$$40B = 12 \times 60$$

$$B = \frac{720}{40}$$

$$B = 18\Omega$$

2. In a Wheatstone bridge, two resistors of 5Ω and RΩ are connected in series with the 5Ω resistor on the left hand side of the bridge. Another two resistors of 3.5Ω and 8Ω are connected in series with the 3.5Ω resistor on the left side. A centre-zero galvanometer connected to the Wheatstone bridge shows a null deflection. Determine the value of R.

Solution

At balance point (null deflection of galvanometer), the relationship between the resistors is given by:

$$\frac{R_1}{R_2} = \frac{R_3}{R_4} \quad (R_1 \text{ and } R_3 \text{ are on the left side of the Wheatstone bridge})$$

Based on the values given in the question, $R_1 = 5\Omega$, $R_2 = R\Omega$, $R_3 = 3.5\Omega$, and $R_4 = 8\Omega$

Therefore, $\dfrac{5}{R} = \dfrac{3.5}{8}$

$$3.5R = 5 \times 8$$

$$R = \frac{40}{3.5}$$

$$R = 11.4\Omega$$

3. If the resistances in a metre bridge are 4Ω and 6Ω, determine the balance point of the metre bridge.

Solution

At balance point, the circuit equation for a metre bridge is given by:

$$\frac{R_1}{R_2} = \frac{l_1}{l_2}$$

$$\frac{4}{6} = \frac{l_1}{100 - l_1} \quad (L_2 = 100 - L_1, \text{ since the metre bridge wire is 1 metre long})$$

$$6L_1 = 4(100 - L_1)$$

$$6L_1 = 400 - 4L_1$$

$$6L_1 + 4L_1 = 400$$
$$10L_1 = 400$$
$$L_1 = \frac{400}{10}$$
$$L_1 = 40cm$$

The balance point is at 40cm

4. A potentiometer is used to compare the emfs of two cells. When a cell of emf 1.5V was used, the balance point was found to be 38cm. What emf will give a balance length of 54cm?

Solution

For a potentiometer, the expression below is valid.

$$\frac{E_1}{E_2} = \frac{l_1}{l_2}$$
$$\frac{1.5}{E_2} = \frac{38}{54}$$
$$38E_2 = 54 \times 1.5$$
$$E_2 = \frac{54 \times 1.5}{38}$$
$$E_2 = 2.13V$$

The emf that will give a balance length of 54cm is 2.13V

5. A potentiometer wire, of resistance 5Ω and length 100cm is connected in series with a 2V cell of internal resistance 0.2Ω. Calculate:

(a) the current flowing through the wire
(b) the balance length for a cell of emf 1.25V

Solution

(a) Recall the emf equation given by:

$$E = I(R + r)$$, and substituting appropriate values gives:
$$2 = I(5 + 0.2)$$
$$2 = 5.2I$$
$$I = \frac{2}{5.2}$$
$$I = 0.38A$$

(b) The voltage across the potentiometer wire is given by:

$$V = IR$$
$$= 0.38 \times 5$$
$$= 1.90V$$

Therefore the relationship between this voltage and the emf of the new cell is given by:

$$\frac{E}{V} = \frac{l}{100}$$
$$\frac{1.25}{1.9} = \frac{l}{100}$$
$$1.9L = 1.25 \times 100$$
$$L = \frac{125}{1.9}$$
$$L = 65.8 \text{cm}$$

The balance length is 65.8cm

6. A potentiometer wire, of resistance 8Ω and length 1m is connected in series with a 4V cell which is also connected in series to a 2Ω resistor. Calculate:
(a) the current flowing through the circuit
(b) the emf of a cell that will give a balance length of 42cm

Solution

(a) The total resistance in the circuit is:

R = 8 + 2   (Since they are in series we add their values)
R = 10Ω

Therefore the current in the circuit is given by:

$$I = \frac{V}{R} \quad \text{(From V = IR)}$$
$$= \frac{4}{10}$$
$$= 0.4A$$

(b) The voltage across the potentiometer wire is given by:

V = IR
= 0.4 x 8   (Use only the resistance of the potentiometer wire)
= 3.2V

Therefore the relationship between this voltage and the emf of the new cell is given by:

$$\frac{E}{V} = \frac{l}{100}$$
$$\frac{E}{3.2} = \frac{42}{100}$$
$$100E = 3.2 \times 42$$
$$E = \frac{134.4}{100}$$
$$E = 1.34V$$

Exercise 17

1. The balance length on a metre bridge for two resistors, X and Y is 60cm on the side of X. If X is 2Ω, calculate the value of Y.

2. In a Wheatstone bridge, two resistors of 10Ω and XΩ are connected in series with the 10Ω resistor on the left hand side of the bridge. Another two resistors of 4.5Ω and 8Ω are connected in series with the 4.5Ω resistor on the left side. A centre-zero galvanometer connected to the Wheatstone bridge shows a null deflection. Determine the value of X.

3. If the resistances in a metre bridge are 5Ω and 6Ω, determine the balance point of the metre bridge.

4. A potentiometer is used to compare the emfs of two cells. When a cell of emf 2V was used, the balance point was found to be 44cm. What emf will give a balance length of 60cm?

5. A potentiometer wire, of resistance 8Ω and length 100cm is connected in series with a 1.5V cell of internal resistance 0.5Ω. Calculate:

(a) the current flowing through the wire

(b) the balance length for a cell of emf 1.1V

6. A potentiometer wire, of resistance 20Ω and length 1.5m is connected in series with a 2V cell which is also connected in series to a 5Ω resistor. Calculate:

(a) the current flowing through the circuit

(b) the emf of a cell that will give a balance length of 40cm

CHAPTER 18
LAWS OF ELECTROLYSIS

Faraday's First Law of Electrolysis

Faraday's first law of electrolysis states that the mass of an element deposited during electrolysis is directly proportional to the product of the current and time.

It is expressed mathematically as follows:

$$m = ZIt$$

where m = mass of element deposited, I = current, t = time in seconds, and Z is a constant called the electrochemical equivalent of the element deposited. The unit of Z is kg/C

Note that, Q = It, where Q is the quantity of charge or electricity.

Faraday's Second Law of Electrolysis

Faraday's second law of electrolysis state that if the same quantity of electricity is passed through different electrolytes, the masses of the elements deposited are proportional to their ratio of relative atomic masses to valency.

It is expressed mathematically as follows:

$$\frac{\text{Mass of element X deposited}}{\text{Mass of element Y deposited}} = \frac{\text{Relative atomic mass / valency of X}}{\text{Relative atomic mass / valency of Y}}$$

Since, $\frac{\text{Relative atomic mass}}{\text{valency}}$ = Chemical equivalent, then:

Faraday's second law of electrolysis can also be expressed mathematically as follows:

$$\frac{\text{Mass of element X deposited}}{\text{Mass of element Y deposited}} = \frac{\text{Chemical equivalent X}}{\text{Chemical equivalent Y}}$$

Note that 1 Faraday (1F or Faraday constant) = 96500C/mol. It is the quantity of electricity needed to library one mole of a monovalent element or half a mole of a divalent element during electrolysis.

In order to determine the mass of an element deposited during electrolysis, and when Faraday constant, F, is given, we use the expression below:

$$m = \frac{MIt}{vF}$$

where m is the mass of element deposited, M is the atomic mass of the element, I is the current, t is the time in seconds, v is the valency of the element, and F is the Faraday constant.

Examples

1. Calculate the time taken to deposit 3.3g of copper using a current of 0.8A in a copper plating process. (Electrochemical equivalent of copper = 0.00033g/C)

<u>Solution</u>

This is Faraday's first law of electrolysis.

Therefore, m = ZIt

$3.3 = 0.00033 \times 0.8 \times t$

$3.3 = 0.000264t$

$t = \dfrac{3.3}{0.000264}$

= 12500 seconds

Or, $t = \left(\dfrac{12500}{60 \times 60}\right)$ hours

t = 3.47 hours (Divide time is seconds by 3600 (i.e. 60 x 60) in order to convert it to hours)

2. Calculate the current needed to deposit 0.112g of silver if the current was passed for 54sec in a silver plating process. (Electrochemical equivalent of silver = 0.00112g/C).
Solution

m = ZIt

$0.112 = 0.00112 \times I \times 54$

$0.112 = 0.06048 I$

$I = \dfrac{0.112}{0.06048}$

= 1.85A

3. If the electrochemical equivalent of silver is 0.00112g/C. how much silver will be deposited in an electrolysis experiment after 9 minutes, by a current of 2.1A?
Solution

m = ZIt

$= 0.00112 \times 2.1 \times (9 \times 60)$ [Note that 9 minutes = (9 x 60) seconds, since the time must be in seconds]

m = 1.27g

Therefore 1.27g of silver will be deposited

4. If 0.033g of copper are deposited in a voltameter when a current of 5A flows for 20sec, calculate the electrochemical equivalent of copper.
Solution

m = ZIt

$0.033 = Z \times 5 \times 20$

$0.033 = 100Z$

$Z = \dfrac{0.033}{100}$

= 0.00033g/C

Therefore the electrochemical equivalent of copper is 0.00033g/C

5. If 0.0066g of copper are deposited in a voltameter when a current of 0.4A flows for 1.5 minutes, how much copper will be deposited after 2 hours by a current of 0.1A?

Solution

There are two cases of experiment here.

For the first electrolysis: $m_1 = ZI_1t_1$ ....Equation 1

For the second electrolysis: $m_2 = ZI_2t_2$ ....Equation 2

Note that Z is constant for both equations since the same metal is involved

Dividing equation 1 by equation 2 gives:

$$\frac{m_1}{m_2} = \frac{I_1 t_1}{I_2 t_2} \quad \text{(Z has cancel out)}$$

$$\frac{0.0066}{m_2} = \frac{0.4 \times (1.5 \times 60)}{0.1 \times (2 \times 60 \times 60)}$$

[Note that 1.5min = (1.5 x 60)sec, and 2hours = (2 x 60 x 60)seconds]

$$\frac{0.0066}{m_2} = \frac{36}{720}$$

$$36 m_2 = 720 \times 0.0066$$

$$m_2 = \frac{4.752}{36}$$

$$m_2 = 0.132g$$

Therefore, 0.132g of copper will be deposited.

Alternatively, this question can be solved by another method by calculating the electrochemical equivalent of copper as follows:

$m = ZIt$

Using the first conditions, we determine Z as follows:

$0.0066 = Z \times 0.4 \times (1.5 \times 60)$

$0.0066 = 36Z$

$$Z = \frac{0.0066}{36}$$

$Z = 0.0001833$

We now use the value of Z to find m in the second conditions as follows:

$m = ZIt$

$\quad = 0.0001833 \times 0.1 \times (2 \times 60 \times 60)$

$m = 0.132g$ (As obtained before)

6. A copper voltameter is connected in series with a silver voltameter, and at the end of a period of time, 5.2g of copper is deposited. Calculate the mass of silver deposited. (Cu = 64, Ag = 108)

Solution

This is a case of Faraday's second law of electrolysis. Hence:

$$\frac{\text{Mass of element X deposited}}{\text{Mass of element Y deposited}} = \frac{\text{Relative atomic mass /valency of X}}{\text{Relative atomic mass /valency of Y}}$$

Let X be copper and Y be silver. The valency of copper is 2, while the valency of silver is 1. Therefore:

$$\frac{\text{Mass of element X deposited}}{\text{Mass of element Y deposited}} = \frac{\text{Relative atomic mass /valency of X}}{\text{Relative atomic mass /valency of Y}}$$

$$\frac{5.2}{Y} = \frac{\frac{64}{2}}{\frac{108}{1}}$$

$$\frac{5.2}{Y} = \frac{32}{108}$$

$$32Y = 108 \times 5.2$$

$$Y = \frac{561.6}{32}$$

$$Y = 17.55g$$

Therefore mass of silver deposited is 17.55g

7. A silver voltameter is connected in series with an aluminium voltameter, and at the end of a period of time, 0.85g of aluminium is deposited. Calculate the mass of silver deposited. (CE of Ag = 108, CE of Al = 9)

Solution

Note that CE represents chemical equivalent.
Therefore, it follows from Faraday's second law that:

$$\frac{\text{Mass of element X deposited}}{\text{Mass of element Y deposited}} = \frac{\text{Chemical equivalent of X}}{\text{Chemical equivalent of Y}}$$

Let X represent silver and Y represent aluminium. Substituting known values gives:

$$\frac{X}{0.85} = \frac{108}{9}$$

$$9X = 108 \times 0.85$$

$$X = \frac{91.8}{9}$$

$$X = 10.2g$$

Therefore mass of silver deposited is 10.2g

8. In the electrolysis of molten sodium chloride, a current of 10A is passed through the electrodes for 2.2 hours. What is the mass of sodium deposited? (1 Faraday = 96500C/mol, Na = 23)

Solution

m = ?, M = 23, I = 10A, t = 2.2hours = (2.2 x 60 x 60)sec, valency of sodium, v = 1, F = 96500

Therefore, $m = \dfrac{MIt}{vF}$ Substituting known vales gives:

$= \dfrac{23 \times 10 \times (2.2 \times 60 \times 60)}{1 \times 96500}$

$= \dfrac{1821600}{96500}$

$m = 18.9g$

9. In an electrolysis experiment, the ammeter records a steady current of 1A. The mass of copper deposited in 28 minutes is 0.67g. Calculate the error in the ammeter reading. (Electrochemical equivalent of copper = 0.00033g/C)

Solution

Let us calculate the accurate value of the current.

$m = ZIt$

Therefore, $I = \dfrac{m}{Zt}$

$= \dfrac{0.67}{0.00033 \times 28 \times 60}$ [Note that 28 minutes = (28 x 60)seconds]

$= \dfrac{0.67}{0.5544}$

$I = 1.21$

Hence, the error in the ammeter reading = Difference between the calculated current and the current given by the ammeter.

$= 1.21 - 1$

Error = 0.21A

10. At the cathode of a copper voltameter, the total surface area of the cathode is $60cm^2$, and a steady current of 4A is maintained in the voltameter for 1 hour. Calculate the thickness of the copper plated on the cathode. (Density of copper = $8.9 \times 10^3 kg/m^3$, electrochemical equivalent of copper = $3.3 \times 10^{-7} kg/C$)

Solution

$m = ZIt$

$= 3.3 \times 10^{-7} \times 4 \times (1 \times 60)$

$= 7.92 \times 10^{-5} kg$

Let us convert the surface in cm^2 to m^2 as follows:

$60cm^2 = (\dfrac{60}{100 \times 100})m^2$ (To convert from cm^2 to m^2, divide by 100^2, i.e. 100 x 100)

$= 6 \times 10^{-3} m^2$

Therefore the surface area is $6 \times 10^{-3} m^2$

Recall that: Density = $\dfrac{mass}{volume}$

Therefore, $8.9 \times 10^3 = \dfrac{7.92 \times 10^{-5}}{V}$

$V = \dfrac{7.92 \times 10^{-5}}{8.9 \times 10^3}$

$V = 8.9 \times 10^{-9} m^3$

Also, Volume = Surface area x thickness   (This is similar to: Volume of a shape = cross-sectional area x height)

$8.9 \times 10^{-9} = 6 \times 10^{-3} \times h$ (where h represents thickness)

$h = \dfrac{8.9 \times 10^{-9}}{6 \times 10^{-3}}$

$h = 1.48 \times 10^{-6} m$

Therefore the thickness of the copper plated on the cathode is $1.48 \times 10^{-6} m$.

Exercise 18

1. Calculate the time taken to deposit 6.6g of copper using a current of 2.6A in a copper plating process. (Electrochemical equivalent of copper = 0.00033g/C)

2. Calculate the current needed to deposit 0.2g of silver if the current was passed for 1.2hours in a silver plating process. (Electrochemical equivalent of silver = 0.00112g/C).

3. If the electrochemical equivalent of silver is 0.00112g/C, how much silver will be deposited in an electrolysis experiment after 2 hours, by a current of 0.5A?

4. If 0.0165g of copper are deposited in a voltameter when a current of 2.1A flows for 3minutes, calculate the electrochemical equivalent of copper.

5. If 0.0132g of copper are deposited in a voltameter when a current of 1A flows for 52 seconds, how much copper will be deposited after 10minutes by a current of 2.4A?

6. A copper voltameter is connected in series with a silver voltameter, and at the end of a period of time, 1.5g of copper is deposited. Calculate the mass of silver deposited. (Cu = 63.5, Ag = 108)

7. A silver voltameter is connected in series with an aluminium voltameter, and at the end of a period of time, 2.1g of aluminium is deposited. Calculate the mass of silver deposited. (CE of Ag = 108, CE of Al = 9)

8. In the electrolysis of molten potassium chloride, a current of 2A is passed through the electrodes for 82minutes. What is the mass of sodium deposited?   (1 Faraday = 96500C/mol, Na = 23)

9. In an electrolysis experiment, the ammeter records a steady current of 1A. The mass of copper deposited in 50 minutes is 0.8g. Calculate the error in the ammeter reading. (Electrochemical equivalent of copper = 0.00033g/C)

10. At the cathode of a copper voltameter, the total surface area of the cathode is $45cm^2$, and a steady current of 5A is maintained in the voltameter for 55 minutes. Calculate the

thickness of the copper plated on the cathode. (Density of copper = $8.9 \times 10^3 \text{kg/m}^3$, electrochemical equivalent of copper = $3.3 \times 10^{-7} \text{kg/C}$)

CHAPTER 19
CONVERSION OF GALVANOMETER TO AMMETER AND VOLTMETER

Conversion of a Galvanometer to an Ammeter

To convert a galvanometer to an ammeter a low resistance shunt, R_S, is connected in parallel with a suitable galvanometer. If the resistance of the galvanometer is R_G, and the current through it is I_G, gender expression relating these terms with the current it is to be converted to, is given by:

$$IR_S = I_G(R_S + R_G),$$

Where I is the current that the galvanometer, is converted to read.
From this equation, if R_S is made the subject of the formula, it gives:

$$R_S = \frac{I_G R_G}{I - I_G}$$

Conversion of a Galvanometer to a Voltmeter

To convert a galvanometer to voltmeter, a high resistance multiplier, R_m is joined in series with a suitable galvanometer. If the resistance of the galvanometer is R_G, and the current through it is I_G, then the p.d or voltage, V, cross the combination is given by:

$$V = I_G(R_m + R_G)$$

Where V is the voltage that the galvanometer is converted to read.
From this equation, if R_m, is made the subject of the formula, it gives:

$$R_m = \frac{V}{I_G} - R_G$$

Examples

1. A galvanometer of internal resistance 10Ω has a full scale deflection with a current of 10mA. Calculate the magnitude of the resistance required to convert it to a voltmeter capable of measuring up to 3V.

<u>Solution</u>

This is a case of conversion of a galvanometer to a voltmeter.
The resistance of the galvanometer, $R_G = 10Ω$
The current through the galvanometer, I_G = 10mA (i.e. 10 milli ampere)

$I_G = 10 \times 10^{-3}$ (Since milli means 10^{-3})

$I_G = 0.01$ (Since $10 \times 10^{-3} = \frac{10}{10^3} = \frac{10}{1000} = 0.01$)

The voltage that the galvanometer is to be converted to, V = 3V
The resistance multiplier required for the conversion, R_m = ? (This is what we are asked to calculate)
The galvanometer equation required for this conversion is given by:

$$V = I_G(R_m + R_G)$$

Therefore, $3 = 0.01(R_m + 10)$
$$3 = 0.01R_m + 0.1$$
$$3 - 0.1 = 0.01R_m$$
$$2.9 = 0.01R_m$$
$$R_m = \frac{2.9}{0.01}$$
$$R_m = 290\Omega$$

Hence the resistance that needs to be connected in series with the galvanometer in order to carry out this conversion is 290Ω.

Alternatively, we can directly use:
$$R_m = \frac{V}{I_G} - R_G$$
$$= \frac{3}{0.01} - 10$$
$$R_m = 300 - 10$$
$$R_m = 290\Omega \quad \text{(As obtained before)}$$

2. A galvanometer gives a full scale deflection of 10mA. If the resistance of the galvanometer is 5Ω, what value of resistance will be used to convert the galvanometer to an ammeter capable of measuring 2A.

Solution

This is a case of conversion of a galvanometer to an ammeter.
The resistance of the galvanometer, $R_G = 5\Omega$
The current through the galvanometer, $I_G = 10mA$ (i.e. 10 milli ampere)
$I_G = 10 \times 10^{-3}$ (Since milli means 10^{-3})
$I_G = 0.01$ (Since $10 \times 10^{-3} = \frac{10}{10^3} = \frac{10}{1000} = 0.01$)

The current that the galvanometer is to be converted to, $I = 2A$
The resistance shunt required for the conversion, $R_s = ?$ (This is what we are asked to calculate)

The galvanometer equation required for this conversion is given by:
$$IR_s = I_G(R_s + R_G)$$
Therefore, $2 \times R_s = 0.01(R_s + 5)$
$$2R_s = 0.01R_s + 0.05$$
$$2R_s - 0.01R_s = 0.05$$
$$1.99R_s = 0.05$$
$$R_s = \frac{0.05}{1.99}$$
$$R_s = 0.0251\Omega$$

Hence the resistance that needs to be connected in parallel with the galvanometer in order to carry out this conversion is 0.0251Ω.

Alternatively, we can directly use:

$$R_S = \frac{I_G R_G}{I - I_G}$$

$$= \frac{0.01 \times 5}{2 - 0.01}$$

$$= \frac{0.05}{1.99}$$

$$= 0.0251Ω \quad \text{(As obtained before)}$$

3. A galvanometer has a full scale deflection of 5mA. If a resistance of 1900Ω is required to convert it to a voltmeter to read 0 - 10V, what is the resistance of the galvanometer?

Solution

This is a case of conversion of a galvanometer to a voltmeter.
The resistance of the galvanometer, R_G = ? (This is what we are asked to calculate)
The current through the galvanometer, I_G = 5mA (i.e. 5 milli ampere)

$$I_G = 5 \times 10^{-3} \text{ (Since milli means } 10^{-3})$$

$$I_G = 0.005 \text{ (Since } 5 \times 10^{-3} = \frac{5}{10^3} = \frac{5}{1000} = 0.005)$$

The voltage that the galvanometer is to be converted to, V = 10V (From a range of 0 - 10V)
The resistance multiplier required for the conversion, R_m = 1900
The galvanometer equation required for this conversion is given by:

$$V = I_G(R_m + R_G)$$

Therefore, 10 = 0.005(1900 + R_G)

10 = 9.5 + 0.005R_G

10 - 9.5 = 0.005RG

0.5 = 0.005R_G

$$R_G = \frac{0.5}{0.005}$$

R_G = 100Ω

Hence the resistance of the galvanometer is 100Ω

4. A galvanometer gives a full scale deflection of 50mA. If a resistance of 0.1Ω is connected in parallel with the galvanometer in order to convert it into an ammeter to read 0 - 3A, what is the resistance of the galvanometer?

Solution

This is a case of conversion of a galvanometer to an ammeter.

The resistance of the galvanometer, R_G = ? (This is what we are asked to calculate)
The current through the galvanometer, I_G = 50mA (i.e. 50 milli ampere)

$I_G = 50 \times 10^{-3}$ (Since milli means 10^{-3})

$I_G = 0.05$ (Since $20 \times 10^{-3} = \frac{50}{10^3} = \frac{50}{1000} = 0.05$)

The current that the galvanometer is to be converted to, I = 3A (From a range of 0 - 3A)
The resistance shunt required for the conversion, R_S = 0.1
The galvanometer equation required for this conversion is given by:

$IR_S = I_G(R_S + R_G)$

Therefore, $3 \times 0.1 = 0.05(0.1 + R_G)$

$0.3 = 0.005 + 0.05R_G$

$0.3 - 0.005 = 0.05R_G$

$0.295 = 0.05R_G$

$R_G = \frac{0.295}{0.05}$

$R_G = 5.9\Omega$

Hence the resistance of the galvanometer is 5.9Ω.

5. An ammeter of resistance 50Ω has a full scale deflection of 20mA. What resistance is needed to convert it into a voltmeter to read 0 - 5V?

Solution

This is a case of conversion of a galvanometer to a voltmeter.
The resistance of the galvanometer, R_G = 50Ω
The current through the galvanometer, I_G = 20mA (i.e. 20 milli ampere)

$I_G = 20 \times 10^{-3}$ (Since milli means 10^{-3})

$I_G = 0.02$ (Since $20 \times 10^{-3} = \frac{20}{10^3} = \frac{20}{1000} = 0.02$)

The voltage that the galvanometer is to be converted to, V = 5V
The resistance multiplier required for the conversion, R_m = ? (This is what we are asked to calculate)
The galvanometer equation required for this conversion is given by:

$V = I_G(R_m + R_G)$

Therefore, $5 = 0.02(R_m + 50)$

$5 = 0.02R_m + 1$

$5 - 1 = 0.02R_m$

$4 = 0.02R_m$

$R_m = \frac{4}{0.02}$

$R_m = 200\Omega$

Hence the resistance that needs to be connected in series with the galvanometer in order to carry out this conversion is 200Ω.

6. A galvanometer of resistance 5Ω has a full scale deflection of 200mA. An XΩ resistor is connected in parallel with the galvanometer to increase the full scale deflection to 1A. What is the value of X?

Solution

This is a case of conversion of a galvanometer to an ammeter.
The resistance of the galvanometer, $R_G = 5Ω$
The current through the galvanometer, $I_G = 200mA$
$I_G = 200 \times 10^{-3}$ (Since milli means 10^{-3})
$I_G = 0.2$
The current that the galvanometer is to be converted to, $I = 1A$
The resistance shunt required for the conversion, $R_S = ?$ (This is X that we are asked to find)
The galvanometer equation required for this conversion is given by:
$IR_S = I_G(R_S + R_G)$
Therefore, we can directly use:
$$R_S = \frac{I_G R_G}{I - I_G}$$
$$= \frac{0.2 \times 5}{1 - 0.2}$$
$$= \frac{1}{0.8}$$
$$= 1.25Ω$$

Hence the resistance that needs to be connected in parallel with the galvanometer in order to convert it to an ammeter is 1.25Ω.

Exercise 19

1. An ammeter of internal resistance 20Ω has a full scale deflection with a current of 100mA. Calculate the magnitude of the resistance required to convert it to a voltmeter capable of measuring up to 3.2V.

2. A galvanometer gives a full scale deflection of 5mA. If the resistance of the galvanometer is 20Ω, what value of resistance will be used to convert the galvanometer to an ammeter capable of measuring 1A.

3. A galvanometer has a full scale deflection of 6mA. If a resistance of 800Ω is required to convert it to a voltmeter to read 0 - 5V, what is the resistance of the galvanometer?

4. A galvanometer gives a full scale deflection of 10mA. If a resistance of 0.04Ω is connected in parallel with the galvanometer in order to convert it into an ammeter to read 0 - 2A, what is the resistance of the galvanometer?

5. An ammeter of resistance 40Ω has a full scale deflection of 80mA. What resistance is needed to convert it into a voltmeter to read 0 - 10V?

6. A galvanometer of resistance 5Ω has a full scale deflection of 100mA. An $R\Omega$ resistor is connected in parallel with the galvanometer to increase the full scale deflection to 3A. What is the value of R?

CHAPTER 20
ALTERNATING CURRENT (A.C) CIRCUIT

The electricity supplied in our homes is not direct current but alternating current. The direct current which gives the same heating rate as a given alternating current is called the effective (or root-mean-square) value of the current. This effective current is given by:

$$I_{r.m.s} = \frac{I_o}{\sqrt{2}}$$

where $I_{r.m.s}$ - root-mean-square (r.m.s) value of the current and I_o = the peak value of the current

Similarly, the effective or rms value of the voltage is:

$$V_{r.m.s} = \frac{V_o}{\sqrt{2}}$$

where V_o = peak value of the voltage

By applying Ohm's law, the voltage and current can be expressed in terms of resistance as follows:

$$V_{rms} = RI_{rms} \quad \text{and} \quad V_o = I_o R$$

where R is the resistance of the circuit.

An a.c circuit may contain one, two or all three of the following components: Resistor, inductor or capacitor. The instantaneous current is the current at a particular time. This current flows through each of the three components in an a.c circuit. It is the same for each component and is given by:

$$I = I_o \sin wt$$

where, I = Instantaneous current, I_o = peak current, and wt = phase angle. Note that w = 2πf.

Examples

1. What is the peak value of a current whose rms value is 7.5A

Solution

$$I_{r.m.s} = \frac{I_o}{\sqrt{2}}$$

$$7.5 = \frac{V_o}{\sqrt{2}}$$

$$I_o = 7.5 \times \sqrt{2}$$

$$= 7.5 \times 1.414$$

$$I_o = 10.6A$$

Therefore, the peak value of the current is 10.6A

2(a). If the mains supply of the voltage in a home is 240V, calculate the peak voltage of the supply.

(b). If this peak voltage has a current of 15A, calculate the value of the effective current.

Solution

(a). The mains supply is the r.m.s value

$$V_{r.m.s} = \frac{V_o}{\sqrt{2}}$$

$$240 = \frac{V_o}{\sqrt{2}}$$

$$V_o = 240 \times \sqrt{2}$$

$$= 240 \times 1.414$$

$$V_o = 339.4V$$

Therefore, the peak value of the voltage is 339.4V

(b). The effective current is the r.m.s value of the current

$$I_{r.m.s} = \frac{V_o}{\sqrt{2}}$$

$$= \frac{15}{\sqrt{2}}$$

$$= \frac{15}{1.414}$$

$$I_{r.m.s} = 10.6A$$

Therefore, the effective value of the current is 10.6A

3. Calculate the instantaneous value of a current, if the mains supply of the current is 15A when its phase angle is 30°.

Solution

$$I = I_o \sin wt$$

But, $I_{r.m.s} = \frac{I_o}{\sqrt{2}}$

Therefore, $I_o = I_{rms} \times \sqrt{2}$

$$= 15 \times 1.414$$

$$I_o = 21.2A$$

Hence, $I = I_o \sin wt$

$$= 21.2 \times \sin 30$$

$$= 21.2 \times 0.5$$

$$I = 10.6A$$

Therefore the instantaneous current is 10.6A

4. The alternating current in a circuit is represented by the equation, $I = 80\sin 200\pi t$, where I is in amperes. Calculate:
(a). The root mean square value of the current
(b). The frequency of the circuit.

Solution
(a). The equation for the instantaneous current in a circuit is: $I = I_o \sin wt$
The equation given in the question is: $I = 80\sin 200\pi t$
Comparing these two equations shows that, $I_o = 80A$

$$\text{But, } I_{rms} = \frac{I_o}{\sqrt{2}}$$
$$= \frac{80}{\sqrt{2}}$$
$$= \frac{80}{1.414}$$
$$I_{rms} = 56.6A$$

Therefore, the room-mean-square (rms) value of the current = 56.6A

(b) $I = 80\sin 200\pi t$
Also, $I = I_o \sin wt$
Comparing these two equations shows that:
$$wt = 200\pi t$$
Hence: $2\pi f t = 200\pi t$
$2f = 200$ (after dividing both sides by πt)
$$f = \frac{200}{2}$$
$f = 100Hz$

5. The current in a circuit at a particular time, t, is given by, $I = 12\sin 350t$. What is:
(a). the current amplitude
(b). The effective value of the current
(c). The frequency of the circuit.

Solution
(a). The equation for the instantaneous current in a circuit is: $I = I_o \sin wt$
The equation given in the question is: $I = 12\sin 350t$
Comparing these two equations shows that, $I_o = 12A$
This value (peak current), is also known as the amplitude value of the current
Therefore the current amplitude is 12A

(b). But, $I_{rms} = \frac{I_o}{\sqrt{2}}$ (I_{rms} represents the effective value of the current)

$$= \frac{12}{\sqrt{2}}$$
$$= \frac{12}{1.414}$$
$$I_{rms} = 8.5A$$

Therefore, the effective (rms) value of the current = 8.5A

(c)   I = 12sin350t
Also,   I = I₀sinwt
Comparing these two equations shows that:
w = 350
Hence: 2πf = 350
$$f = \frac{350}{2\pi}$$
$$= \frac{350}{2 \times 3.142}$$
$$= \frac{350}{6.248}$$
f = 55.7Hz

Exercise 20
1. What is the peak value of a current whose rms value is 12A
2(a). If the mains supply of the voltage in a home is 220V, calculate the peak voltage of the supply.
(b). If the peak voltage in (a) above, has a current of 10A, calculate the value of the effective current.
3. Calculate the instantaneous value of a current, if the mains supply of the current is 5A when its phase angle is 60°.
4. The alternating current in a circuit is represented by the equation, I = 100sin140πt, where I is in amperes. Calculate:
(a). The r.m.s value of the current
(b). The frequency of the circuit.
5. The current in a circuit at a particular time, t, is given by, I = 0.6sin120t. What is:
(a). the current amplitude
(b). The effective value of the current
(c). The frequency of the circuit.

CHAPTER 21
RESISTOR, INDUCTOR AND CAPACITOR (R-L-C) CIRCUIT IN SERIES

Reactance in a.c circuit

In a circuit containing an inductor or a capacitor, there is opposition to the flow of current by these components. The opposition to the flow of current by the inductor or capacitor is called reactance.

Inductive reactance

This is defined as the opposition offered to the flow of current in an ac circuit through the inductor. It is denoted by XL and measured in Ω. It is given by:

$$X_L = \omega L \quad \text{or} \quad X_L = 2\pi f L$$

where w = angular velocity, f = frequency and L = inductance of the inductance. The unit of inductance is the Henry (H).

The voltage across the inductor is given by:

$$V_L = IX_L,$$

where I is the current in the circuit.

Capacitive reactance

This is defined as opposition given to the flow of current in an a.c circuit through the capacitor. It is denoted by X_C and measured in Ω. It is given by:

$$X_C = \frac{1}{\omega C} \quad \text{or} \quad X_C = \frac{1}{2\pi f C}$$

where C = capacitance of the capacitor.

The voltage across the capacitor is given by:

$$V_C = IX_C$$

Resistor and Inductor (R-L) circuit

In this type of circuit, the effective voltage in the circuit is given by:

$$V = \sqrt{V_R^2 + V_L^2}$$

where, V_R is the voltage across the resistor, and V_L is the voltage across the inductor.

Note that the voltage across the resistor is given by:

$$V_R = IR$$

The total opposition given to the current flow across the circuit is called impedance. Impedance is defined as the total opposition given to the flow of current in an a.c circuit by a resisto and either an inductor or a capacitor or both. It is denoted by Z and measured in Ω.

For an R-L circuit, the impedance is given by:

$$Z = \sqrt{R^2 + X_L^2} \quad \text{or} \quad Z = \frac{V}{I}$$

The phase angle for this type of circuit is given by:

$$\tan\theta = \frac{X_L}{R} \quad \text{or} \quad \tan\theta = \frac{V_L}{V_R}$$

where θ is the phase angle

Resistor and Capacitor (R-C) circuit

In an R-C circuit, the effective voltage in the circuit is given by:

$$V = \sqrt{V_R^2 + V_C^2}$$

where, V_R is the voltage across the resistor, and V_C is the voltage across the capacitor.

For an R-C circuit, the impedance is given by:

$$Z = \sqrt{R^2 + X_C^2} \quad \text{or} \quad Z = \frac{V}{I}$$

The phase angle for this type of circuit is given by:

$$\tan\theta = \frac{X_C}{R} \quad \text{or} \quad \tan\theta = \frac{V_C}{V_R}$$

Inductor and Capacitor (L-C) circuit

In an L-C circuit, the effective voltage is given by:

$$V = V_L - V_C$$

The effective reactance/overall reactance or impedance in this type of circuit is given by:

$$Z = X_L - X_C$$

Resistor, Inductor and Capacitor (R-L-C) circuit

In an R-L-C circuit, the effective voltage in the circuit is given by:

$$V = \sqrt{V_R^2 + (V_L - V_C)^2}$$

where, V_R is the voltage across the resistor, V_L is the voltage across the inductor, and V_C is the voltage across the capacitor.

For an R-L-C circuit, the impedance is given by:

$$Z = \sqrt{R^2 + (X_L - X_C)^2}$$

Or, $\quad Z = \frac{V}{I}$

The phase angle for this type of circuit is given by:

$$\tan\theta = \frac{X_L - X_C}{R} \quad \text{or} \quad \tan\theta = \frac{V_L - V_C}{V_R}$$

Note that for all the circuit types impedance is given by: $Z = \frac{V}{I}$, thus, $V = IZ$

Resonance in an a.c circuit

Resonance is said to occur when the maximum current in the circuit is attained. At resonance the total reactance is zero, and Z = R. This occurs when $X_L = X_C$, i.e. $\omega L = \dfrac{1}{\omega C}$

Or, $2\pi f_o L = \dfrac{1}{2\pi f_o C}$

Making f_o the subject of the formula above gives:

$$f_o = \dfrac{1}{2\pi \sqrt{LC}}$$

where, f_o is the resonant frequency.

Vector diagram

In R-L-C circuit, V_R is in phase with the current, I. The V_L leads the current by 90° (or $\dfrac{\pi}{2}$ rad.), while V_C lags behind the current, by 90°. Thus V_L and V_C are exactly 180° out of phase.

Power in a.c circuit

Instantaneous power, P, consumed in an a.c circuit is given by:

P = IV

where I and V are the instantaneous values.

The average power consumed in an a.c circuit is given by:

$P_{av} = \dfrac{I_o V_o CosQ}{2}$ or $P_{av} = IVcos\,\theta$

where I and V are effective values (or r.m.s values), and $cos\,\theta$ = power factor. The power factor is given by: $Cos\,\theta = \dfrac{R}{Z}$

The power lost or dissipated in the circuit is given by:

P = I²R

Note that in a.c circuits, the voltages given are usually rms voltages, even if they are not specified as rms values.

Also, the values of power dissipated and average power consumed are equal.

This means that I²R and IVcos θ will give the same values.

Examples

1. A wire has an inductance of 18mH and is connected to a voltage source of 240V, having a frequency of 50Hz. Calculate:
(a) the inductive reactance
(b) the current flowing in the wire

Solution
(a) The inductance, L = 18mH = 18 x 10⁻³H (Since milli means 10^{-3})

Therefore, $L = \dfrac{18}{10^3} = \dfrac{18}{1000}$

L = 0.018H

The inductive reactance is given by:

$X_L = 2\pi f L$

$= 2 \times \dfrac{22}{7} \times 50 \times 0.018$

$X_L = 5.66\Omega$

Therefore the inductive reactance is 5.66Ω

(b) Recall that: $V = IX_L$

Therefore, $I = \dfrac{V}{X_L}$

$= \dfrac{240}{5.66}$

I = 42.4A

The current in the circuit is 42.4A

2. A capacitor has a capacitance of 3μF, and is connected to a current source of 0.2A and a frequency of 60Hz. Calculate:
(a) the capacitive reactance
(b) the voltage across the capacitor

Solutions
(a) The capacitance, C = 3μF = 3 x 10⁻⁶F (Since micro, μ, means 10^{-6})

Therefore, $C = \dfrac{3}{10^6} = \dfrac{3}{1000000}$

C = 0.000003F

The capacitive reactance is given by:

$X_C = \dfrac{1}{2\pi f C}$

$= \dfrac{1}{2 \times \dfrac{22}{7} \times 50 \times 0.000003}$

$= \dfrac{7}{2 \times 22 \times 50 \times 0.000003}$

(By the rule of division, 7 can be moved to become the numerator. 3.142 can also be used for the value of π)

$X_C = 1060.6\Omega$

Therefore the capacitive reactance is 1060.6Ω

(b) Recall that: $V = IX_C$

= 0.2 x 1060.6

V = 212V

The voltage across the capacitor is 212V

3. A coil of inductance, 0.007H, a resistor of resistance 6Ω and a capacitor of capacitance 0.002F are connected in series to an ac source of frequency (500/π)Hz. If the rms voltage across the coil, the resistor and capacitor are 30V, 30V and 70V respectively, calculate:
(a) the rms voltage of the source
(b) the impedance
(c) the rms current in the circuit
(d) power dissipated
(e) write down the equation for the rms voltage, V, in terms of the time, t.

Solution

(a) Given: V_R = 30V, V_L = 30V, V_C = 70V

The voltage of the source is the effective voltage in the circuit. It is given by:

$$V = \sqrt{V_R^2 + (V_L - V_C)^2}$$
$$= \sqrt{30^2 + (30 - 70)^2}$$
$$= \sqrt{900 + (-40)^2}$$
$$= \sqrt{900 + 1600}$$
$$= \sqrt{2500}$$

V = 50V

The rms voltage of the source is 50V

(b) We need to calculate the capacitive and inductive reactance in order to calculate the impedance

Let us calculate the inductive reactance as follows:

The inductance, L = 0.007H

The inductive reactance is given by:

$X_L = 2\pi f L$

$= 2 \times \pi \times \dfrac{500}{\pi} \times 0.007$ (There is no need of writing π as $\dfrac{22}{7}$ since f is given in terms of π)

$= 2 \times 500 \times 0.007$ (The π has cancelled out)

$X_L = 7Ω$

Let us calculate the capacitive reactance as follows:

The capacitance, C = 0.002F

The capacitive reactance is given by:

$$X_C = \frac{1}{2\pi f C}$$

$$= \frac{1}{2 \times \pi \times \frac{500}{\pi} \times 0.002}$$

$$= \frac{1}{2 \times 500 \times 0.002}$$

$$= \frac{1}{2}$$

$$X_C = 0.5\Omega$$

The impedance is given by:

$$Z = \sqrt{R^2 + (X_L - X_C)^2}$$
$$= \sqrt{6^2 + 7 - 0.5)^2}$$
$$= \sqrt{36 + (6.5)^2}$$
$$= \sqrt{36 + 42.25}$$
$$= \sqrt{78.25}$$

$$Z = 8.8\Omega$$

The impedance in the circuit is 8.8Ω

(c) Current, I, is given by:

$$I = \frac{V}{Z}$$

$$= \frac{50}{8.8}$$ [Note that the voltage in the circuit (i.e the effective voltage) should be used)]

I = 5.7A

The rms current in the circuit is 5.7A

(d) The power dissipated is given by:

P = I²R
= 8.8² x 6
= 8.8 x 8.8 x 6
= 77.44 x 6
P = 465W

Therefore the power dissipated is given by 465W

(e) The equation for voltage (instantaneous voltage) is given by:

V = V₀sinwt
V = V₀sin2πft

But, V₀ = V$_{rms}$√2   (Since V$_{rms}$ = $\frac{V_o}{\sqrt{2}}$)

= 50 x √2

= 50 x 1.414
V₀ = 70.7
Therefore, V = V₀sin2πft

$$= 70.7\sin(2 \times \pi \times \frac{500}{\pi} \times t)$$

= 70.7sin(2 x 500) x t   (The π cancels out)

V = 70.7sin100t

Therefore, the equation for the rms voltage, V, in terms of the time, t is, V = 70.7sin100t

4. An a.c circuit consists of an inductor of 0.8H and a resistor of 100Ω. The voltage applied is 240V and its frequency is 50Hz. Find:
(a) the inductive reactance
(b) the impedance
(c) the current in the circuit
(d) the voltage across the inductor
(e) the voltage across the resistor
(f) the phase angle between the voltage and current
(g) the power factor
(h) the average power supplied

Solutions

(a) The inductance, L = 0.8H
The inductive reactance is given by:

$X_L = 2\pi f L$

$= 2 \times \frac{22}{7} \times 50 \times 0.8$

$= \frac{1760}{7}$

$X_L = 251\Omega$

Therefore the inductive reactance is 251Ω

(b) The impedance is given by:

$Z = \sqrt{R^2 + X_L^2}$

$= \sqrt{100^2 + 251^2}$

$= \sqrt{10000 + 63001}$

$= \sqrt{73001}$

$= 270\Omega$

The impedance in the circuit is 270Ω

(c) The current is given by:

$$I = \frac{V}{Z}$$
$$= \frac{240}{270}$$ (Note that the voltage usually given in the question is the rms voltage, unless otherwise stated)

I = 0.89A (Note that this current is the r.m.s current, since the voltage is the r.m.s voltage)

The current in the circuit is 0.89A

(d) The voltage across the inductor is given by:
$$V = IX_L$$
$$= 0.89 \times 251$$
$$= 223V$$
The voltage across the inductor is 223V

(e) The voltage across the resistor is given by:
$$V = IR$$
$$= 0.89 \times 100$$
$$V = 89V$$
Therefore the voltage across the resistor is 89V

(f) The phase angle between the voltage and current in an R-L circuity is given by:
$$\tan\theta = \frac{X_L}{R}$$
$$= \frac{251}{100}$$
$$\tan\theta = 2.51$$
Therefore, $\theta = \tan^{-1}2.51$
$$\theta = 68.3°$$
The phase angle between the voltage and current is 68.3°

(g) The power factor is given by:
$$\cos\theta = \frac{R}{Z}$$
$$= \frac{100}{270}$$
$\cos\theta = 0.37$ (Note that this can also be obtained by finding the cosine of the phase angle, i.e. cos 68.3 = 0.37)

The power factor is 0.37

(h) The average power supplied is given by:

$P_{av} = IV\cos\theta$

$= 0.89 \times 240 \times \dfrac{100}{270}$ ($\cos\theta = \dfrac{R}{Z}$, which is $\dfrac{100}{270}$)

$= \dfrac{21360}{270}$

$= 79.1W$

Therefore the average power supplied is 79.1W

5. An a.c circuit consist of a capacitor of reactance 110Ω. If the frequency of the circuit is 60Hz, determine the capacitance of the capacitor.

Solution
The capacitive reactance is given by:

$X_C = \dfrac{1}{2\pi f C}$

$110 = \dfrac{1}{2 \times \dfrac{22}{7} \times 60 \times C}$

$110 = \dfrac{7}{2 \times 22 \times 60 \times C}$

$110 = \dfrac{7}{2640\,C}$

$110 \times 2640C = 7$

$290400C = 7$

$C = \dfrac{7}{290400}$

$C = 0.000024F$

$= 24 \times 10^{-6}F$

$C = 24\mu F$ (Since μ = micro = 10^{-6})

Hence the capacitance is 24μF

6. An a.c circuit consist of an inductor of reactance 62Ω. If the frequency of the circuit is 50Hz, determine the inductance of the inductor.

Solution
The inductive reactance is 62Ω, and it is given by:

$X_L = 2\pi f L$

$62 = 2 \times \dfrac{22}{7} \times 50 \times L$

$62 = \dfrac{2200 L}{7}$

$2200L = 62 \times 7$

$L = \dfrac{434}{2200}$

$L = 0.20H$

Therefore the inductance of the inductor is 0.20H

7. A 60µF capacitor in series with a 40Ω resistor is connected to a 100V, 50Hz, a.c supply. Calculate:
(a) the impedance
(b) the current in the circuit
(c) the potential difference across the capacitor
(d) the phase angle between the voltage and current
(e) the power factor
(f) the average power supplied
(g) the power dissipated in the circuit

Solutions

(a) In order to calculate the impedance, we must first calculate the capacitive reactance. This is as shown below:

The capacitance, $C = 60\mu F = 60 \times 10^{-6} = \frac{60}{10^6} = 0.000060F$

The capacitive reactance is given by:
$$X_C = \frac{1}{2\pi f C}$$
$$= \frac{1}{2 \times \frac{22}{7} \times 50 \times 0.000060}$$
$$= \frac{7}{2 \times 22 \times 50 \times 0.000060}$$
$$= \frac{7}{0.132}$$
$$X_C = 53\Omega$$

The impedance is given by:
$$Z = \sqrt{R^2 + X_C^2}$$
$$= \sqrt{40^2 + 53^2}$$
$$= \sqrt{1600 + 2809}$$
$$= \sqrt{4409}$$
$$Z = 66\Omega$$

The impedance in the circuit is 66Ω

(b) The current is given by:
$$I = \frac{V}{Z}$$
$$= \frac{100}{66}$$

I = 1.51A (Note that this current is the r.m.s current, since the voltage is the r.m.s voltage)

Therefore the current in the circuit is 1.51A

(c) The potential difference across the capacitor is given by:
$$V = IX_C$$
$$= 1.51 \times 53$$
$$= 80V$$
The potential difference across the capacitor is 80V

(d) The phase angle between the voltage and current in an R-C circuit is given by:
$$\tan\theta = \frac{X_C}{R}$$
$$= \frac{53}{40}$$
$$\tan\theta = 1.325$$
Therefore, $\theta = \tan^{-1} 1.325$
$$\theta = 53°$$
The phase angle between the voltage and current is 53°

(e) The power factor is given by:
$$\cos\theta = \frac{R}{Z}$$
$$= \frac{40}{66}$$
$\cos\theta = 0.60$ (Note that this can also be obtained by finding the cosine of the phase angle, i.e. cos 53 = 0.60)

Hence the power factor is 0.60

(f) The average power supplied is given by:
$$P_{av} = IV\cos\theta$$
$$= 1.51 \times 100 \times \frac{40}{66} \quad (\cos\theta = \frac{R}{Z}, \text{ which is } \frac{40}{66})$$
$$= \frac{6040}{66}$$
$$= 91.5W$$
Therefore the average power supplied is 91.5W

(g) The power dissipated in the circuit is given by:
$$P = I^2 R$$

= 1.51² x 40
= 1,51 x 1.51 x 40
= 91.2W

8. A series R-L-C circuit is connected to a 60Hz, a.c mains voltage of 220V. If the resistance is 50Ω, capacitance is 8µF and inductance is 0.6H, calculate:
(a) the capacitive and inductive reactance
(b) the current in the circuit
(c) the phase angle between the voltage and current
(d) the power factor
(e) the power dissipated in the resistor
(f) the resonant frequency

Solution
(a) The inductive reactance is calculated as follows:
The inductance, L = 0.6H
The inductive reactance is given by:

$X_L = 2\pi f L$

$= 2 \times \dfrac{22}{7} \times 60 \times 0.6$

$= \dfrac{2 \times 22 \times 60 \times 0.6}{7}$

$= \dfrac{1584}{7}$

$X_L = 226\Omega$

Therefore, the inductive reactance is 226Ω
Let us calculate the capacitive reactance as follows:

The capacitance, $C = 8\mu F = 8 \times 10^{-6} F = \dfrac{8}{10^6} = 0.000008F$

The capacitive reactance is given by:

$X_C = \dfrac{1}{2\pi f C}$

$= \dfrac{1}{2 \times \dfrac{22}{7} \times 60 \times 0.000008}$

$= \dfrac{7}{2 \times 22 \times 60 \times 0.000008}$ (By the rule of division in fraction, the 7 can be moved up to become the numerator)

$= \dfrac{7}{0.02112}$

$X_C = 331\Omega$

(b) In order to determine the current in the circuit, we have to first calculate the impedance. The impedance is given by:

$$Z = \sqrt{R^2 + (X_L - X_C)^2}$$
$$= \sqrt{50^2 + (226 - 331)^2}$$
$$= \sqrt{2500 + (-105)^2}$$
$$= \sqrt{2500 + 11025}$$
$$= \sqrt{13525}$$
$$Z = 116\Omega$$

The impedance in the circuit is 116Ω

Therefore, the current, I, is given by:

$$I = \frac{V}{Z}$$
$$= \frac{220}{116}$$
$$I = 1.90A$$

The current in the circuit is 1.90A

Note that this current is the rms current since the voltage is taken to be the rms voltage

(c) The phase angle between the voltage and current in an R-L-C circuit is given by:

$$\tan \theta = \frac{X_L - X_C}{R}$$
$$= \frac{226 - 331}{50}$$
$$= \frac{-105}{50}$$
$$\tan \theta = -2.1$$

Therefore, $\theta = \tan^{-1}(-2.1)$
$$\theta = -64.5°$$

The phase angle between the voltage and current is −64.5°. This means that the voltage lags behind the current by 64.5°.

(d) The power factor is given by:

$$\cos \theta = \frac{R}{Z}$$
$$= \frac{50}{116}$$
$$\cos \theta = 0.431 \quad \text{(This can also be obtained from the phase angle as, cos64.5 = 0.431)}$$

(e) The power dissipated in the resistor is given by:

$$P = I^2R$$
$$= 1.90^2 \times 50$$
$$= 1.90 \times 1.90 \times 50$$
$$= 3.61 \times 50$$
$$P = 180.5W$$

Therefore the power dissipated in the resistor is 180.5W

(f) The resonant frequency is given by:
$$f_o = \frac{1}{2\pi\sqrt{LC}}$$
$$= \frac{1}{2 \times 3.142 \times \sqrt{0.6 \times 0.000008}} \quad \text{(The value of } \pi \text{ has been taken to be 3.142)}$$
$$= \frac{1}{2 \times 3.142 \times \sqrt{0.0000048}}$$
$$= \frac{1}{6.284 \times 0.00219}$$
$$= \frac{1}{0.013762}$$
$$= 72.7 \text{ Hz}$$

The resonant frequency is 72.7Hz

9. A source of emf 240V and frequency 50Hz is connected to an R-L-C circuit in series. The potential difference across the resistor is 140V and that across the inductor is 50V when the current across the capacitor is 10A. Calculate the:
(a) potential difference across the capacitor
(b) the capacitance of the capacitor
(c) the resistance of the resistor
(d) the inductance of the inductor

Solutions

(a) Recall that the voltage source in an R-L-C circuit is given by:
$$V = \sqrt{V_R^2 + (V_L - V_C)^2} \quad \text{Substituting known values from the question gives:}$$
$$240 = \sqrt{140^2 + (50 - V_C)^2} \quad \text{We will now find } V_C \text{ which is the voltage across the capacitor.}$$
$$240 = \sqrt{19600 + (50 - V_C)^2}$$

Squaring both sides of the equation gives:
$$57600 = 19600 + (50 - V_C)^2$$
$$57600 - 19600 = (50 - V_C)^2$$
$$(50 - V_C)^2 = 38000$$

Taking the square root of both sides gives:
$$50 - V_C = \pm 195$$

Since 50 - V_C (i.e. V_L - V_C) gives a large value of 195, it shows that V_C must be greater than V_L. If this is the case, then we have to take the negative square root.

Therefore, 50 - V_C = - 195

$\qquad$ 50 + 195 = V_C

$\qquad$ V_C = 245V

Therefore the voltage across the capacitor is 245V

(b) In order to determine the capacitance of the capacitor, we have to first find the capacitive reactance.

Recall that the voltage across the capacitor is given by:

$\qquad$ V_C = IX_C

$\qquad$ 245 = 10 x X_C

$\qquad$ $X_C = \dfrac{245}{10}$

$\qquad$ X_C = 24.5Ω

But capacitive reactance is given by:

$\qquad$ $X_C = \dfrac{1}{2\pi f C}$

$\qquad$ $24.5 = \dfrac{1}{2 \times 3.142 \times 50 \times C}$

$\qquad$ $24.5 = \dfrac{1}{314.2C}$

$\qquad$ 24.5 x 314.2C = 1

$\qquad$ 7698C = 1

$\qquad$ $C = \dfrac{1}{7698}$

$\qquad$ C = 0.000130F

Or, $\qquad$ C = 130μF $\qquad$ (When divided by 10^{-6})

The capacitance of the capacitor is 130μF

(c) The voltage across the resistor is given by:

$\qquad$ V_R = IR

$\qquad$ 140 = 10R

$\qquad$ $R = \dfrac{140}{10}$ $\qquad$ (Note that the same current of 10A flows across all three circuit components since they are in series)

$\qquad$ R = 14Ω

Therefore the resistance of the resistor is 14Ω

(d) In order to determine the inductance of the inductor, we have to first find the inductive reactance. Recall that the voltage across the inductor is given by:

$$V_L = IX_L$$
$$50 = 10 \times X_L$$
$$X_L = \frac{50}{10}$$
$$X_L = 5\Omega$$

But inductive reactance is given by:
$$X_L = 2\pi f L$$
$$5 = 2 \times 3.142 \times 50 \times L$$
$$5 = 314.2L$$
$$L = \frac{5}{314.2}$$
$$L = 0.0159$$

Therefore, the inductance of the inductor is 0.0159H

Exercise 21

INSTRUCTIONS: In all questions where necessary, use 3.142 for the value of π.

1. A wire has an inductance of 20mH and is connected to a voltage source of 220V, having a frequency of 50Hz. Calculate:
(a) the inductive reactance
(b) the current flowing in the wire

2. A capacitor has a capacitance of 10µF, and is connected to a current source of 0.5A and a frequency of 50Hz. Calculate:
(a) the capacitive reactance
(b) the voltage across the capacitor

3. A coil of inductance, 0.02H, a resistor of resistance 14Ω and a capacitor of capacitance 0.0005F are connected in series to an ac source of frequency $(200/\pi)$Hz. If the rms voltage across the coil, the resistor and capacitor are 80V, 80V and 20V respectively, calculate:
(a) the rms voltage of the source
(b) the impedance
(c) the rms current in the circuit
(d) power dissipated
(e) write down the equation for the rms voltage, V, in terms of the time, t.

4. An a.c circuit consist of an inductor of 2H and a resistor of 100Ω. The voltage applied is 110V and its frequency is 50Hz. Find:
(a) the inductive reactance
(b) the impedance
(c) the current in the circuit
(d) the voltage across the inductor

(e) the voltage across the resistor
(f) the phase angle between the voltage and current
(g) the power factor
(h) the average power supplied

5. An a.c circuit consist of a capacitor of reactance 200Ω. If the frequency of the circuit is 55Hz, determine the capacitance of the capacitor.

6. An a.c circuit consist of an inductor of reactance 105Ω. If the frequency of the circuit is 60Hz, determine the inductance of the inductor.

7. A 100μF capacitor in series with a 20Ω resistor is connected to a 210V, 50Hz, a.c supply. Calculate:
(a) the impedance
(b) the current in the circuit
(c) the potential difference across the capacitor
(d) the phase angle between the voltage and current
(e) the power factor
(f) the average power supplied
(g) the power dissipated in the circuit

8. A series R-L-C circuit is connected to a 50Hz, a.c mains voltage of 240V. If the resistance is 80Ω, capacitance is 20μF and inductance is 0.025H, calculate:
(a) the capacitive and inductive reactance
(b) the current in the circuit
(c) the phase angle between the voltage and current
(d) the power factor
(e) the power dissipated in the resistor
(f) the resonant frequency

9. A source of emf 220V and frequency 50Hz is connected to an R-L-C circuit in series. The potential difference across the resistor is 100V and that across the inductor is 70V when the current across the capacitor is 5A. Calculate the:
(a) potential difference across the capacitor
(b) the capacitance of the capacitor
(c) the resistance of the resistor
(d) the inductance of the inductor

10. The average power consumed in an a.c. circuit is 100W. If the impedance of the circuit is 110Ω and the current in the circuit is 2A, calculate the:
(a) power factor
(b) the phase angle between the current and voltage in the circuit.

ANSWERS TO EXERCISES

Exercise 1
(1)(a) 3.14sec (b) 0.318Hz (c) 0.2m/s (d) 0.4m/s^2 (e) 0.173m/s
(2)(a) 1.99sec (b) 0.316m/s (3)(a) 0.72m/s (b) 1.02m/s (4) 0.71Hz
(5) 6sec (6)(a) 0.02sec (b) 18.85m/s(middle). The velocity at the end is zero
(7)(a) 2.97m (b) 14.96m/s^2 (8) 28.3sec (9) 7.2sec (10) 4.08rad/sec
(11)(a) 10rad/sec (b) 0.98m/s
(12)(a) A = 10m, f = 1Hz, T = 1sec (b) $v = -62.84\sin(2\pi t + \frac{\pi}{4})$, $a = -394.9\cos(2\pi t + \frac{\pi}{4})$
(c) x = 7.07m, v = −44.4m/s, a = −279.2m/s^2
(13)(a) 32m/s (b) 128m/s^2 (c) 10.2m (14) 6sec (15) 1 : 4

Exercise 2
(1)(a) 0.281sec (b) 0.05J (c) 0.05J (2)(a) 0.397sec (b) 0.0225J
(3)(a) K.E = 3.2 x 10^{-6}J, P.E = = 4 x 10^{-7}J (b) = 3.6 x 10^{-6}J
(4)(a) 8.1cm (b) P.E = 0.00197J, K.E = 0.0032J (c) 0.00517J
(5) 0.00216J (6) 0.000222J (7)(a) K.E = 2.49 x 10^{-6}J, P.E = 4.74 x 10^{-7}J
(b) 2.96 x 10^{-6}J (8)(a) 0.562sec (b) 0.002J (c) 0.002J

Exercise 3
(1) 2m/s^2 (2) 40Hz (3) 0.21rad/sec (4)(a) 0.167sec (b) 37.6rad/sec
(c) 7.5m/s (d) 282.8m/s^2 (e) 1414s^{-2} (5)(a) 0.786rad/sec
(b) 0.628m/s (c) 0.125Hz (d) 0.492m/s^2 (6) 12.38N (7) 0.123N
(8)(a) 0.97rad/sec (b) 0.116m/s (c) 0.00113N

Exercise 4
(1)(a) A = 200cm (or 2m), λ = 160cm (or 1.6m) (b) 1.26sec (c) 1.27m/s
(2) 3.2m (3) 0.375m (4) 0.46Hz (5) 0.0533sec
(6)(a) 10m (b) 4Hz (c) 50m/s (d) 12.5m (e) 0.503m^{-1}
(7)(a) 20m (b) 0.01sec (c) 500m/s (d) $y = 8\sin 2\pi(25t + \frac{x}{20})$
(8)(a) 4m (b) 0.318Hz (c) 3.14sec (d) 0.40m/s (e) 2.0rad/sec
(f) 1.26m (g) 5.0m^{-1} (9)(a) 0.07m (b) 0.2Hz (c) 0.24m/s
(10)(a) 1m (b) 1.25m/s (c) $y = 1.5\sin 2\pi(1.25t + x)$

Exercise 5
(1) 660m (2) 7.06sec (3) 332m/s (4)(a) 1.71sec (b) 2.35sec
(5) 49 (6) 324m/s (7) 338m/s (8) 378m/s

Exercise 6
(1)(a) 4Hz (b) 0.25sec (2)(a) 1.25Hz (b) 53.25Hz (3) 6Hz
(4) 23Hz and 17Hz (5)(a) 0.71Hz (b) 324.3Hz

Exercise 7
(1) 196.4Hz (2) λ = 45cm, f = 1511Hz (3) λ = 0.2m, f = 1020Hz
(4) 1320Hz (5) 8.05m (6) 364m/s (7) 333.8m/s (8) 354.2Hz

Exercise 8
(1) 68.5Hz (2) 100.6Hz (3) 8.5N (4) 111.8Hz (5) 297Hz
(6) 62Hz (7)(a) 41.4Hz (b) 69Hz (c) 58m/s (8) 55Hz
(9) 14.4Hz (10) 50cm

Exercise 9
(1) 135Hz (2) 0.88rec/sec (3) 37 teeth (4) 22.2sec (5) 152sec

Exercise 10
(1) 232.3Hz (2) 357.3Hz (3) 450.2Hz (4) 375Hz (5) 460Hz
(6) 347Hz (7) 398.4Hz (8) 441.4Hz (9) 300.7m/s (10) 3.6m/s

Exercise 11
(1) 50A (2) 36.7Ω 3.(a) 3.2 x $10^{-4}\Omega$m (b) 3125Ω^{-1}m^{-1} (4) 2 x 10^{-5}sec
(5) 3A 6.(a) 240$(\Omega m)^{-1}$ (b) 4.17 x $10^{-3}\Omega$m

Exercise 12
(1) 8.7Ω (2) 0.48Ω (3) 6.67Ω (4) Emf = 4.5V, r = 3.6Ω
(5) Emf = 2V, r = 0.8Ω

Exercise 13
1.(a) 1A (b) 5v and 9V (c) 14V
2.(a) 1.09A (b) 2.18V, 4.36V, 5.45V respectively
3.(a) 2.625A across 1Ω and 0.875A across 3Ω (b) 2.625V (c) 2.625V
4.(a) 10A across 5Ω, 6.25A across 8Ω and 5A across 10Ω (b) 50V (c) 21.25A
5.(a) 1.8A (b) 0.6A across each 6Ω, 1.44A across 1Ω and 0.36A across 4Ω
(c) 3.6V across each 6Ω, 9V across 5Ω and 1.44V across 1Ω and 4Ω
6.(a) 11V (b) 5Ω 7.(a) 1.25A across 8Ω, 2.5A across RΩ and 5a across 2Ω (b) 4Ω

Exercise 14
1.(a) 0.255A (b) 0.255V (2) 6.5A 3.(a) 0.124A (b) 0.0124V
(4) 0.75Ω (5) 0.36A and 0.72V (6) 2.5Ω

Exercise 15
(1) 864kJ   2.(a) 0.25A   (b) 960Ω   (3) 43,200J
4.(a) 44.1W   (b) 3,810,240J or 3.81MJ   5.(a) 0.0631A   (b) 3484.8Ω
6.(a) 61,440W   (b) 737,280J

Exercise 16
(1) 69.7cents   (2) $45.02   (3) 4.05cents   (4) 66.1cents   (5) $13.85

Exercise 17
(1) 1.33Ω   (2) 17.8Ω   (3) 45.5cm on the 5Ω side   (4) 2.73V
5.(a) 0.176A   (b) 78cm   6.(a) 0.08A   (b) 0.43V

Exercise 18
(1) 7,692sec   (2) 0.0413A   (3) 4.032g   (4) 4.37×10^{-5}g/C   (5) 0.366g
(6) 5.1g   (7) 25.2g   (8) 2.35g   (9) 0.19A   (10) 1.36×10^{-4}m or 0.0136cm

Exercise 19
(1) 12Ω   (2) 0.1Ω   (3) 33Ω   (4) 9.96Ω   (5) 85Ω   (6) 0.172Ω

Exercise 20
(1) 17.0A   2.(a) 311V   (b) 7.1A   (3) 6.15A   4.(a) 70.7A   (b) 70Hz
5.(a) 0.6A   (b) 0.42A   (c) 19.1Hz

Exercise 21
1.(a) 6.28Ω   (b) 35A   2.(a) 318.3Ω   (b) 159.1V
3.(a) 100V   (b) 14.32Ω   (c) 6.98A   (d) 682W   (e) V = 100sin400t
4.(a) 628.4Ω   (b) 636.3Ω   (c) 0.173A   (d) 108.7V   (e) 17.3V
(f) 81.0°   (g) 0.1572   (h) 3.0W   (5) 1.45μF   (6) 0.28H
7.(a) 37.59Ω   (b) 5.59A   (c) 178.3V   (d) 57.9°   (e) 0.532
(f) 625W   (g) 625W   8.(a) 159Ω and 7.86Ω   (b) 1.40A   (c) −62.1°
(d) 0.468   (e) 156.8W   (f) 225Hz
9.(a) 266V   (b) 5.98×10^{-5}F or 59.8μF   (c) 20Ω   (d) 0.0446H
10.(a) 0.25   (b) 75.5

www.ingramcontent.com/pod-product-compliance
Lightning Source LLC
Chambersburg PA
CBHW082108220526
45472CB00009B/2095